Mechatronic Systems

Mechatronic Systems
Modelling and Simulation with HDLs

Georg Pelz

Infineon Technologies, Munich, Germany

Translated by

Rachel Waddington

Member of the Institute of Translation and Interpreting

WILEY

Other Wiley Editorial Offices

John Wiley & Sons Inc., 111 River Street, Hoboken, NJ 07030, USA

Jossey-Bass, 989 Market Street, San Francisco, CA 94103-1741, USA

Wiley-VCH Verlag GmbH, Boschstr. 12, D-69469 Weinheim, Germany

John Wiley & Sons Australia Ltd, 33 Park Road, Milton, Queensland 4064, Australia

John Wiley & Sons (Asia) Pte Ltd, 2 Clementi Loop #02-01, Jin Xing Distripark, Singapore 129809

John Wiley & Sons Canada Ltd, 22 Worcester Road, Etobicoke, Ontario, Canada M9W 1L1

Library of Congress Cataloging-in-Publication Data

Pelz, Georg, 1962-
   [Modellierung und Simulation mechatronischer Systeme. English]
   Mechatronic systems : modelling and simulation with HDLs / George Pelz.
   p. cm.
   Includes bibliographical references and index.
   ISBN 0-470-84979-7 (alk. paper)
   1. Mechatronics. 2. Computer hardware description languages. I. Title.

TJ163.12.P4513 2003
621–dc21

2002192433

British Library Cataloguing in Publication Data

A catalogue record for this book is available from the British Library

ISBN 0-470-84979-7

Typeset in 10.5/13pt Times by Laserwords Private Limited, Chennai, India
Printed and bound in Great Britain by Antony Rowe Ltd, Chippenham, Wiltshire
This book is printed on acid-free paper responsibly manufactured from sustainable forestry
in which at least two trees are planted for each one used for paper production.

Contents

Preface

Most of this work came into being during my employment at the Chair for Electron Devices and Circuits in the Electronics Engineering department of the Gerhard-Mercator University, Duisburg. Section 7.5 covers material that I have worked on for my current employer, Infineon Technologies.

At this point I would like to express my gratitude for the support that I received from many sides. My special thanks go to Prof. Dr. G. Zimmer, in whose department I was able to work continuously for many years on the subject of this book, and who helped me in many ways in the process. Moreover, I would like to thank Prof. Dr. M. Glesner for his support of the work.

I would also like to thank my colleagues at the Gerhard-Mercator University, Duisburg, the Fraunhofer Institut IMS and Infineon Technologies, who provided a great deal of assistance in the form of discussions and suggestions during the preparation of the book. The following in particular should be mentioned: Dr. J. Bielefeld, Dr. M. Leineweber, Dipl.-Ing. A. Lüdecke and Dipl.-Ing. L. Voßkämper.

Apart from the technical side, I would like to express my thanks to Tilmann Leopold. Last, but not least, I thank my family for their encouragement and support during the composition of this book.

Ebersberg, January 2003   Georg Pelz (Georg.Pelz@onlinehome.de)

1

Objective and Motivation

1.1 Introduction

The objective of this work was to support the design of mechatronic systems by the use of simulations. This raises the question of what exactly is mechatronics. Current definitions describe mechatronics as an interaction between electronics, mechanics and information technology, see Isermann [164] or Wallaschek [421]. It makes no difference here whether we are talking about macromechanics or micromechanics. In the former case we speak of mechatronics, in the latter of micromechatronics or microelectromechanical systems (MEMS). As was discovered during the course of this project, although the dimensions of the mechanics in the systems under investigation may vary, the methods used for modelling and simulation are largely the same, which makes the joint consideration of macromechanics and micromechanics an obvious approach.

Why is the modelling and simulation of mechatronic systems difficult? First of all, the field of mechatronics incorporates very different domains and similarly varied methods of description. The field of electronics includes analogue and digital, as well as continuous and event-oriented, processes. The same is true of mechanics, although often for totally different reasons. In the field of mechanics, events may, for example, be triggered by the transition from static to sliding friction. In electronics, on the other hand, an event is brought about by the flicking of a switch, triggering a connection to the entire digital world. In mechanics we also have to deal with geometric aspects in three spatial dimensions. Furthermore, multibody and continuum mechanics of different representational forms also have to be taken into account. Finally, software can be considered as information in bistable circuits and thus classified as electronics. However, this is not sufficient to achieve an efficient and transparent consideration, which means that we have to develop our own models for the software.

The development of models is thus a difficult process at the best of times and one which is prone to errors. However, a systematic verification and validation of the model is not in sight. As in other fields of simulation, models containing errors can produce arbitrary results. Recognising such errors is often not a simple matter.

Mechatronic Systems   Georg Pelz
© 2003 John Wiley & Sons, Ltd   ISBN: 0-470-84979-7

This is particularly true if the simulation relates to the design of a technical system and its task is to make predictions about the system's functionality. In this case the system in question does not exist at all in the real world, which means that no measurements are available for checking the model. Rather, the design has yet to be investigated and completed. So proving the correctness of a model is a matter of importance. If we now interpret — as did Butterfield in [55] — a model as a scientific theory, then the validation of the model must be placed within narrow boundaries. According to Popper [338] the following is true for the validation of a theory:

> *In order to be scientific, a theory must be falsifiable. It must be empirically testable, at least in principle, and there must be a test that disproves the theory in the event of a negative outcome.*
>
> *There can never be a rigorous validation of a scientific theory. The best that we can do is to develop empirical tests for the theory — fair tests, but the stricter the better — and to hold onto the theory only as long as it has passed all tests.*

The same applies for the validation of models. We can develop as many tests for a model as we like, but this does not prove the validity of the model. At best, trust in a model increases with the number of tests.

Depending upon the problem to be solved, we can differentiate between two fundamental starting points in the simulation of mechatronic systems. If the mechanical part of a mechatronic system is to be developed, then the mechanics should be developed taking into account the electronics. In this case electronics and software are commonly considered as a regulatory function and dealt with along with the mechanics in the form of suitable equations. The purpose of this work is to investigate the opposite case — the development of electronics and software taking into account the mechanical component. This type of design should be supported by simulations.

Hardware description languages, which have been widespread in the field of electronics for some time, and for which various commercial simulators are already available, represent the tools for achieving this end. Anything that can be modelled using a hardware description language can also be simulated.

Thus the task is primarily a modelling problem. Furthermore, standards exist for hardware description languages, which means that models can be exchanged between simulators. One example is the IEEE standard VHDL 1076.1 (VHDL-AMS) [160], which permits the description of digital and analogue systems. The aim of this work is to cover the entire breadth of modelling for mechatronic and micromechatronic systems using hardware description languages and to thereby take a direct route to the corresponding simulations.

This structure of this work is as follows: After the introduction, the second chapter deals with the principles of modelling and simulation for electronics and mechanics. Particular importance is attributed to the verification and validation of models. The third chapter describes state of the art techniques for the simulation of mechatronics and micromechatronics. Chapter 4 supplies the most important constructs of digital and analogue hardware description languages. Chapters 5 and 6 deal comprehensively with the methods for the consideration of software and mechanics in hardware description languages. This creates a compendium of basic methods that can be combined at will according to the system under consideration. This is illustrated in Chapters 7 and 8 on the basis of six demonstrators for mechatronics and micromechatronics. The ninth chapter finally summarises the work and highlights its most important conclusions. At the end of the book there is a bibliography, the appendix containing lists of symbols, trademarks, and abbreviations used, plus the index.

2

Principles of Modelling and Simulation

2.1 Introduction

The introduction of Information Technology in the last fifty years has allowed modelling and simulation to penetrate the majority of engineering disciplines and natural and social sciences. Regardless of whether the matter under debate is the design of wheel suspension for a car, the metabolism of a bacteria, or the introduction of a new interest formula, models of these real systems are always drawn upon to gain an understanding of the inner relationships of the system and to make predictions about its behaviour. The simulation is often also used as a substitute for experiments on an existing system, which is associated with a range of benefits:

- In comparison to real experiments, virtual experiments often require a significantly lower outlay in financial terms and in terms of time, because it is generally considerably cheaper to model virtual prototypes than it is to build real prototypes.

- Some system states cannot be brought about in the real system, or at least not in a non-destructive manner.

- Normally all aspects of virtual experiments are repeatable, something that either cannot be guaranteed for the real system or would involve considerable cost.

- Simulated models are generally completely controllable. So all input variables and parameters of the system can be predetermined. This is normally not the case for a real system.

- Simulated models are generally fully monitorable. All output variables and internal states are available, whereas in the real system every variable to be monitored involves at least a significant measurement cost. In addition, each measurement taken influences the behaviour of the system.

Mechatronic Systems   Georg Pelz
© 2003 John Wiley & Sons, Ltd   ISBN: 0-470-84979-7

- In some cases the 'time constants' of the experiment and observer are incompatible, such as the investigation of elementary particles or galaxies.

- In some cases an experiment is ruled out for moral reasons, for example experiments on humans in the field of medical technology.

However, these benefits are countered by some disadvantages:

- Each virtual experiment requires a complete, validated and verified modelling of the system.

- The accuracy with which details are reproduced and the simulation speed of the models is limited by the power of the computer used for the simulation.

In many cases the benefits outweigh the disadvantages and virtual experiments can be used advantageously. The repeatability guaranteed by the computer is particularly beneficial if the virtual experiment is systematically planned and performed as part of an optimisation.

In what follows we will define a range of terms relating to modelling and simulation. This will allow us to move from a general consideration to the systems investigated in this work, thus providing a good structure to the discussion. The following representation relates to the work of the SCS Technical Committee on Model Credibility, see [362].

Reality is initially an entity, situation or system to be investigated by simulation. Its modelling can be viewed as a two-stage process, as shown in Figure 2.1. In the first stage, reality is *analysed* and modelled using verbal descriptions, equations, relationships or laws of nature, which initially establishes a *conceptual* model. A field of application then has to be defined for this conceptual model, within which the model should provide an acceptable representation of reality. Furthermore, the degree of correspondence between conceptual model and reality that should be achieved for the selected field of application, has to be defined. A conceptual model is adequately *qualified* for a predetermined field of application if it produces the required degree of correspondence with reality. In the second stage of modelling the conceptual model is transformed into an executable, i.e. *simulatable*, model as part of *implementation*. This primarily consists of a set of instructions that describe the system's response to external stimuli. The instructions can be processed manually or using a computer. The latter is called *simulation* and permits the processing of significantly greater data quantities, and thus the consideration of significantly more complex problems.

The development of models for simulation is a difficult process, and thus prone to errors. On the other hand, the reliability of a simulation is crucially dependent upon the quality of the model. So methods and tools are required that are capable of validating and verifying the models. Let us now define these two terms, *validation* and *verification*, more closely, see Figure 2.1. Model verification investigates whether the executable model reflects the conceptual model within the specified limits of accuracy. Verification transfers the conceptual model's field of application

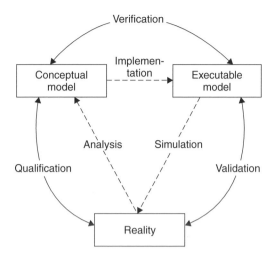

Figure 2.1   Model generation, simulation, validation and verification in context

to the executable model. Model validation, on the other hand, should tell us whether the executable model is suitable for fulfilling the envisaged task within its field of application. In other words: Verification ensures the system is modelled *right*, whereas validation is all about modelling the *right* system. Various degrees of validity can be defined for a model:

Replicative validity

A model is replicatively valid if it moves along tracks that have already been marked out by measurements upon the real system. This is the lowest level of validity. Such models may, for example, be used in the field of training to teach people to use a real system by means of virtual experiments.

Predictive validity

A model is predictively valid if it 'predicts' data that are not extracted from the system until later. So, for example, simulations supply important information on the functionality of a circuit even before it has been constructed in the form of a chip or board. It is also perfectly possible to mix predictively valid component models with replicatively valid models if measurement data is available for the modelling of some components but not for others. A predictively valid model is also replicatively valid.

Structural validity

A model is structurally valid if it not only describes the outward behaviour of a real system accurately enough, but also imitates the internal processes for the generation of the behaviour at the pins. This is the highest level of validity and this level in particular is required in order to understand the real system. A structurally valid system is also predictively valid.

2.2 Model Categories

We can obtain an initial classification of models by considering the range of values of the system variables, see for example Zeigler [435]. These may be continuous or discrete. A range of values is *continuous* if it covers real numbers or an interval of them. For example, a mechanical position has a continuous range of values. In a *discrete* range of values, on the other hand, the system variable takes on a value from a finite (or at least countable) quantity of values, as is the case for digital, electronic signals. The states of the model take on a discrete, continuous or mixed form depending upon the system variables.

Time is explicitly removed from the system variables and investigated in a similar manner with respect to its value range. In the *discrete* case time proceeds in leaps; valid time points are calculated as the product of a whole number and a basic time span. This may, for example, be suitable if a gate simulation is run with unit delays. By contrast, we can also consider models in which time is *continuous*. These can be divided into two categories: *event-oriented* models and *differential equation* models. In the former case each change of state of the model is triggered by an event, so that the trajectory of system states proceeds in leaps. The events themselves can occur at arbitrary points in time; their number in relation to a predetermined time interval is however finite. By contrast, in models based upon differential equations the trajectory of system states is continuous. Changes are described on the basis of the system variables and their rate of change.

A further possibility for differentiating between models is based upon whether the description uses *concentrated* or *distributed* parameters. Examples of the former case are electronic components or the fixed and elastic bodies of the multibody representation of a mechanical system. Distributed parameters should be used in the consideration of a mechanical continuum, for example.

Models may furthermore be of a *static* or *dynamic* nature. In the former case, in electronics for example, when determining the operating point of a circuit it is sufficient to represent capacitors as open circuits and coils as short-circuits. In multibody mechanics stationary systems can be analysed. Dynamic models are required in electronics for transient simulations, i.e. for those over a time range, whereas in mechanics we can differentiate between two application cases: kinematics and kinetics, see for example Nikravesh [299]. Kinematics relates to the investigation of positions, speeds and accelerations without taking into account the forces that cause the movement they describe. Kinetics also considers the acting forces.

In some cases a model cannot be described in a purely deterministic manner, meaning that at least one random variable must be included. As an example, a model may serve to evaluate the power of a computer, which accesses its hard drive with a probability of x% and its tape deck with a probability of y%. Models containing at least one random variable are classified as *stochastic*. All others are called *deterministic*.

A further option for the classification of models is the consideration of the 'outside world' of a model. If the model is isolated from the outside world and thus has no inputs and outputs, then it is called *autonomous*. All other models are called *non-autonomous*. An autonomous model produces a movement in the state space from itself, without taking in and producing data, whereas a non-autonomous model primarily converts values at the inputs into the outputs based upon the current state.

A final option for the classification of models is represented by the question of whether or not time crops up explicitly in the model equations. In the former case the model is *time-variant*, in the latter *time-invariant*.

2.3 Fields of Application

2.3.1 Introduction

If technical systems are to be developed, two main fields of application can be identified for the simulation: The validation of specifications and the verification of designs. In the ideal case the specification or design will be available immediately in model form, so that nothing stands in the way of direct simulation. Hitherto this has mainly been the case in the design of digital electronics using hardware description languages. Otherwise, modelling must take place first to bring about the transition from an arbitrary description to a simulatable model.

The use of modelling and simulation is closely linked to the underlying design processes. These can be roughly divided in accordance with their design direction into top-down and bottom-up design flows. In what follows these will be briefly introduced and characterised by their influence upon modelling.

2.3.2 Bottom-up design

Bottom-up design is the classic method of development of electronics and mechanics, see Figure 2.2. The initial starting point is a specification, which is typically drawn up in natural language. Then the basic components, e.g. transistors, resistors, capacitors or springs, masses, shock absorbers, joints, etc. are added and combined successively to form ever more complex and abstract creations until a complete design emerges. This takes place on a structural level, so that the only thing that is determined each time is which submodules make up a module and how these are

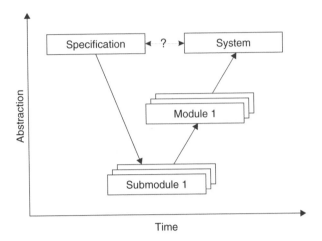

Figure 2.2   Bottom-up design process

to be connected together. Such a design can be performed using a circuit editor or a suitable tool for multibody systems.

The primary advantage of bottom-up design is that the influences of a nonideal implementation can be taken into account at an early stage. For electronics these may be unavoidable parasitic resistances, capacitances and inductances. In the field of mechanics they may be friction effects, for example.

However, one problematic aspect is coming upon the specification for the design, after having had to take a 'diversion' via the submodules and modules from the abstract functional description. This is because, as a result of the structure-oriented modelling, a system can only be simulated when it has been completely implemented. Thus errors and weaknesses in the system design are not noticed until a late stage, which can bring about considerable costs and delays.

2.3.3 Top-down design

A significant characteristic of top-down design is the prevailing design direction from abstract to detailed descriptions, see Figure 2.3. The starting point is a pure behavioural model, the function of which already covers a good part of the specification. The model is successively partitioned and refined until an implementation is obtained. It is necessary to describe a system or module of it in a functional manner. This was first made possible by the introduction of hardware description languages in the field of electronics. Using these the design is directly formulated as a model, so that most of the modelling can be dispensed with.

The top-down design sequence has the following advantages:

- Errors and weaknesses in the design are noticed early, in contrast to the bottom-up approach.

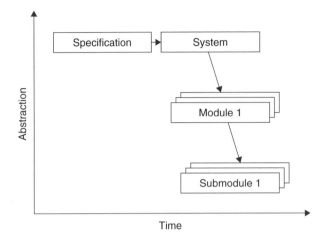

Figure 2.3 Top-down design sequence

- The implementable part of the specification can be validated by simulations.

- The implementable part of the specification is available as a precisely defined reference for the verification of the design.

- The functional part of the specification is unambiguous and complete (in contrast to a specification in natural language). In the event of doubt, a simulation is run.

- The implementable specification and the models of the individual design stages mean that full documentation is available, which however still remains to be supplemented by comprehensive commentary.

In the case of mixed-signal design, the implementable specification can be made available to the test engineers at an early stage as part of a 'simultaneous engineering' approach. Using a model for the testing machine a virtual test is created, in which test programmes can be developed on the workstation. This removes the fixed sequence of design → production → test development and also saves a great deal of time on test development.

However, the disadvantage of the use of implementable specifications is that some technical content can be expressed in a simpler, more compact and more easily understood form in natural language than in a formal modelling language. In addition, there is the question of adhering to the formally correct description of the desired semantics, which incurs an additional cost in relation to a paper specification. Finally, problems in the physical realisation, such as excessive delay times for certain blocks, are not recognised until a relatively late stage.

For mechanics the top-down design sequence is still in the development stage. A significant reason for this is that unified and standardised description methods for mechanical behaviour, with which a design can be taken incrementally from an abstract specification to a detailed implementation, are only now being developed.

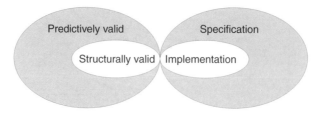

Figure 2.4   Level of validity and its significance for the design of a technical system

2.3.4   Relationship of design strategies to modelling

In the case of the top-down design sequence, modelling is used for the specification of the desired behaviour or for the formulation of designs. In both cases the result can be directly checked through simulation; there is no such thing as modelling exclusively for the purpose of simulation. In this connection, an important classification of such models by their level of validity can be made, see Figure 2.4. For a specification, predictive validity is sufficient — the manner in which the terminal behaviour of the specified systems and modules is individually generated is not relevant. A system design, on the other hand, ideally supplies a structurally valid model that describes both the terminal behaviour and the inner structure.

By contrast, if a technical system is to be developed using a bottom-up design sequence, then simulation can be used for checking the system design or parts of it after the conclusion of the design phase. Modelling is thus not an integral part of the design process; instead it is often performed exclusively for the purpose of the simulation, which raises questions regarding the verification and validation of the model.

Where modelling is used outside a design process we can differentiate between the following two cases: structurally valid modelling in natural and social sciences in order to gain understanding of a system; and replicatively valid modelling in the field of training. The former plays only a lesser role in the consideration of technical systems. The latter is used primarily for the imitation of familiar behaviour. A well-known example is flight simulators that are used for the training of pilots in all feasible operational situations. Such simulators are now available on the market for almost all types of vehicle. But simulators can also be used for other types of training. Preparation for the repair of the Hubble telescope involved a great deal of expenditure on simulation due to the considerable costs and the narrow time frame for such measures in space, see Loftin [237] and [242].

2.3.5   Modelling for the specification

The main purpose of a specification is to describe the desired behaviour of a system to be developed and the associated boundary conditions. Classically, a specification is available on paper, which is associated with a whole range of problems.

First of all it raises the question of its validity, i.e. whether the described system really corresponds with the desired system. Furthermore, it is doubtful whether a given (paper) specification is completely and unambiguously formulated. These questions can only be answered in a systematic manner when the transition is made to an implementable specification, which can then be validated by simulation, for example. A further advantage of this transition lies in the possibility of the verification of the individual design stages against the specification. Furthermore, this opens up the opportunity of performing a formal verification against the specification. In digital electronics, behavioural modelling as a specification is becoming increasingly prevalent, in all other domains it is still at a very early stage.

Modelling for a specification is pure behavioural modelling, which — as is the case for a paper specification — may not anticipate the implementation. For a microprocessor, for example, a specification would describe only the instruction set and the associated actions. The way that the individual operations are realised cannot be the object of the specification. An executable specification for a memory module may consist of a large array for the memory content and some logic for the processing of read and write processes. The specification of an A/D converter could formulate the pure translation of analogue values into digital values and the resulting delay.

2.3.6 Modelling for the design

Modelling for the checking of technical system designs for each simulation is the classic application case. All engineering-science disciplines use simulation beneficially to this end.

This applies particularly in microelectronics. A manufacturing run typically lasts 6–12 weeks and is associated with significant costs. Repairs to manufactured chips are more or less impossible. Under such boundary conditions, one cannot afford to iterate the manufacturing process to rectify design errors. Instead, it is necessary to enter manufacture with a fundamentally error-free design, which — given the complexities that are currently possible, involving some tens of millions of transistors — cannot be achieved without simulation.

If we consider discretely structured printed circuit boards, then it is slightly less critical that the circuit is fully checked in advance by simulation. The etching and fitting of circuit boards is significantly simpler and quicker than chip manufacture. Changes can be performed comparatively easily. The circuits are also less complex by orders of magnitude. So it can be worthwhile to solder a circuit together as a bread-board arrangement and check it by measurement. Nevertheless, the performance of virtual experiments on a computer is generally quicker and cheaper than the real experiment in the laboratory.

For software, things are comparatively simple. The compilation of software can be regarded as rudimentary modelling, as software is executable after this stage, i.e. it is simulatable. The simulation sequence and the simulation result are normally displayed in a debugger that shows the current status of the software, i.e. program line and variable values, plus their outputs on the terminal. Without this type of simulation, software development would be unthinkable.

Like electronics, the construction of mechanical systems in reality is very expensive in terms of time and costs. In many of the industries in question the answer to this problem lies in the increased use of simulation. The automotive industry is particularly advanced in this field. The two main key words here are *digital mock-up* and *virtual prototype*, see for example Paulini *et al.* [317] or Schweer *et al.* [376]. A *digital mock-up* is as complete as possible a description of a single product on the computer and thus represents a limited data quantity. All the various tools check the design on the basis of this data. The digital mock-up thus primarily represents a medium for information exchange, which links together data sources and data sinks in the design process. At regular intervals, for example every two weeks [376], new data are put in and thus are available to all possible users. A *virtual prototype* is extracted from the data of the digital mock-up, which can then be used for experiments on the computer. A classic example of this is the simulation of crash tests. In this application, a finite-element model is obtained from the CAD data of the body by automatic meshing, which can then be subjected to any desired crash situations. Although the simulation requires several hours of processing time even on the fastest computer, it means that the majority of real crash tests can be dispensed with. Furthermore, simulations are also run in virtually all other sectors of the automotive industry, such as for example in the development of running gear, engine, drive train and the associated electronics.

2.4 Model Development

2.4.1 Introduction

The following section provides an overview of the most up-to-date methods for model development in electronics and mechanics, looking at both the common ground and differences. We can make an initial classification by asking whether the model describes the structure or the behaviour of a system.

Taking the first case, in classic modelling the model establishes only which components make up the system and how these are connected together. Alternatively, however, the term structural modelling can also be expanded and, for example, take in the description of the structure of an equation system or a finite state machine. In such cases the following forms of model description may be called structural: electronic circuit diagrams, state graphs, multibody diagrams, meshes of finite elements, block diagrams, bond graphs and Petri nets. The common factor of all these descriptive forms is that they are all graphical in nature.

If, on the other hand, it is the behaviour of a system that is to be described then this can be achieved on the basis of the underlying physics or the measured input/output behaviour. In the former case the development of such models is relatively costly and requires a comprehensive understanding of the system. On the other hand, such models can be adapted to the actual system over a wide range by modifying parameters. If, for example, a system is to be driven by a DC motor, various makes can be included in the simulation by the use of the applicable parameters. These 'generic' models thus cover a whole class of components. As an alternative to modelling on the basis of physical behaviour the other option is to take measured data and feed this into models. This is also called experimental modelling and is used if physical modelling is not implementable or the resulting model is too complex for the desired purpose. Typically, however, experimental modelling has to be repeated every time one of the components in question is altered. Both in the case of physical and experimental modelling the models are generally formulated on the basis of equations and assignments, i.e. consequently formulated in the form of text.

In addition to a simulation, an emulation may also come into consideration under certain speed requirements. This has different characteristics for electronics and mechanics. In the field of digital electronics the term emulator is used to mean a device that can take on the function of any desired digital circuit, see for example Bender and Kaiser [25]. This function is based upon a number of programmable chips, for example so-called FPGAs, the logic functions of which are stored in a local RAM and can thus be modified. Currently up to a hundred thousand gate functions can be stored on a single FPGA. With regard to speed, FPGAs, and thus emulators, are generally significantly slower than dedicated hardware, but are, however, faster than a simulation by orders of magnitude. The emulation of analogue electronics and mechanics on the other hand is based upon signal processors, so-called DSPs, that are optimised for analogue signal processing, see for example Huang *et al.* [155] or Georgiew [116]. So differential equation models of mechanical components can again be calculated faster than is the case for a simulator by orders of magnitude.

Since modelling is a difficult process, and prone to errors, in some cases real components are embedded into a simulation, see for example Helldörfer *et al.* [136] or Le *et al.* [219]. This is also called 'hardware in the loop'. This does not mean that the entire system is constructed as an electronic bread-board assembly or mechanical prototype, instead usually just one component is fitted. Alternatively, the environment of the system to be developed can be included in real form. The rest of the system is modelled in the classical manner, so that simulated and real behaviour are mixed together. The advantage of this is that the modelling and its validation can be dispensed with for the real hardware in the simulation loop. However, the principle disadvantage is that the real components have to be fully installed in the laboratory and adequately fitted with actuators and sensors in order to ensure the main inputs and outputs. Furthermore, the simulation of the remainder of the system must in this case take place in real time, which may involve considerable cost, depending upon the system. Alternatively, this real time simulation can be replaced by an emulation to speed things up.

All the methods described up to this point relate to the description of an error-free system. This is worthwhile if the simulation is to contribute to the actual design. In some cases, however, the aim is to investigate the effect of errors within the system. In this case error modelling is called for. One application for this is the evaluation of measures to increase intrinsic safety; another is the evaluation of test methods for differentiating between functional systems and rejects during production. In both cases, errors that impair the function of the system under consideration are modelled. Here too the modelling represents an abstraction of reality, which in the ideal case covers several error mechanisms. For example, the stuck-at error model in digital electronics describes the permanent presence of a logical 0 or logical 1 at a signal of the circuit. Whether this is caused by a short-circuit with a supply cable or by excessively deep etching of contact holes is of secondary importance. The decisive point is that the circuit no longer functions correctly and that this problem can be detected by the tests developed.

Due to their importance, structural, physical and experimental model development will be considered in more depth in the following. Finally, we note that specialist fields, such as modelling with neural networks, fuzzy techniques or genetic programming, will not be considered.

2.4.2 **Structural modelling**

Introduction

A structural model is characterised by the basic models used and the connection structure between these basic models. A module can be composed of basic models and can itself be again connected to other modules. This can be performed successively, thus describing complex systems. A structural model can be characterised on the basis of the following terms: Hierarchy, modularity, regularity and locality. The hierarchy of a model is derived from the call structure of basic models and modules. So an operational amplifier (=module) can be put together from MOS transistors (=basic models) and then circuits can be built up from operational amplifiers. Using graph theory, such a hierarchy can be described as a tree, in which the roots represent the system as a whole and the leaves represent the basic models. The number of levels of the hierarchy grow in a logarithmic relationship to the number of basic elements involved. The modularity of the system relates to the question of how simple and reasonable it is to divide the system into modules. Regularity is a measure of how many module types are necessary to represent the entire system. A low number is beneficial here because it indicates a compact representation. Finally, locality is a measure of how well a module can be considered without the context of its installation. Modules with straightforward interfaces to their outside world are particularly beneficial here.

In the following, models are considered in the form of circuit diagrams, state graphs, multibody diagrams and finite elements. Further descriptions with structural aspects are bond graphs, block diagrams and Pr/T networks.[1] As these descriptive forms also permit a modelling of electro-mechanical systems, these are described in detail in Chapter 3 as alternatives to modelling using hardware description languages.

Circuit diagrams

In the case of design using a circuit diagram editor, modelling is primarily used for the derivation of a net list, which is used as a circuit model, incorporating the component or gate models. This procedure is so simple and unproblematic that the process of modelling a circuit is not generally perceived as such. Likewise, there are not normally any problems with the validation of the circuit model. In the most extreme case there may be verification problems with the program for deriving the net list. The field of application is predominantly the development of analogue circuits. Although digital circuits can also be developed using circuit diagrams, a top-down design process is only possible using behavioural modelling based upon hardware description languages.

State graphs

Digital systems can also be represented by state graphs with the system structure then being stored on relatively abstract levels. The selection of the state transitions is precisely specified by conditions. Furthermore, in state graphs only the structure of the connections is necessary in order to characterise the model in question. Such a model can, for example, be used for the specification of digital behaviour, but it can also be translated into a programming or hardware description language and then used directly for the design of software and hardware.

Multibody diagrams

Things are more complicated for multibody mechanics. Although the importance of structural modelling is gaining increasing recognition here too, see for example the work of Panreck [313], when drawing up the model equations it is often the system as a whole that is considered rather than viewing it as a collection of components. Only with the introduction of object-oriented modelling, see Otter [308] or Kecskeméthy [185], does the structural modelling of multibody systems also become more prevalent.

Finite elements

A particularly graphic form of structural modelling is to break down mechanical structures into finite elements for the modelling of continuum mechanics. This is

---

[1] Predicate/transition network.

also called meshing, and both geometric dimensions and topological information are important. The element matrices of the individual finite elements are found from their material parameters and geometry, whereas the connection structure between the elements, and consequently the system matrix, is derived from the topology. Often the meshing has to be checked manually in order to ensure that the elements have the correct form, the grid is sufficiently fine and available symmetries are exploited.

2.4.3   Physical modelling

Introduction

In physical modelling the laws of physics are used to describe the behaviour and inner action mechanism of a system or a component. The selection of the relevant relationships depending upon suitability and efficiency and the establishment of cause and effect chains, requires a comprehensive understanding of the system and remains an engineering task. Computer support for this form of modelling is at best rudimentary.

In the following, some classifications will be undertaken for the characterisation of the physical modelling based upon various criteria. These consider the perspectives of modelling and the nature of the yielded equations. Otherwise the reader is referred at this point to Chapters 5 and 6 on modelling, and also to Chapters 7 and 8 on applications, which contain a whole range of examples of physical modelling and electro-mechanical systems.

Perspectives of modelling

The perspectives of modelling offer a coarse division of the physical models which, however, runs through all disciplines like a red thread. We should differentiate here between whether the system perspective or the component perspective has been selected. In one case the system-oriented modelling formulates the system in the overall context; in the other case object-oriented modelling describes components, which only form a system by their connection together, i.e. by structural modelling. The decisive factor is that in object-oriented modelling no system knowledge is fed into the component model. This ensures that the components can be used in any desired context, so that modelling work only has to be performed once and not for each system.

Hitherto in electronics, more significance has been attached to object-oriented modelling. The physical models for electronic components provide the classic example of this. These are formulated independently of the circuit in which they are used. The connection structure is determined in a circuit diagram, which forms a structural model. Thus the validation of the circuit model is in principle achieved by a validation of the component model. This is particularly worthwhile if the number of basic models is small. But object-orientation is also becoming increasingly prevalent in digital design using hardware description languages, although in this context it should be regarded more in the context of an increase in efficiency in the development of text-based, software-like models, see for example Ecker and Mrva [93].

In mechanics object-orientation has only recently been implemented in order to make modelling easier, whereby the work of Otter [308] and Kecskeméthy [185] in particular, are worth mentioning. One explanation for this is the fact that the number of basic elements and the associated variation in mechanics is significantly greater than is the case in electronics. Furthermore, the classic modelling methods of mechanical engineering often lead to descriptions in the form of generalised coordinates,[2] which are again incompatible with object-oriented modelling. The advantage of the generalised coordinates is that the resulting equation system has a minimum number of equations and, furthermore, the constraints can be disregarded for holonomous systems. This is attractive from a numerical point of view. However, generalised coordinates can only be specified by drawing upon knowledge of the entire system and not from the mole-hill perspective of a component.

Resulting equations

In this section we will investigate the equations that result from the various modelling forms. From a mathematical point of view, a digital gate or the setting of a digital signal in a hardware description language gives an instruction, which is executed after the passage of a predetermined time period. This period corresponds with the time delay of the described block. If the block is defined without a delay, then a virtual period of time still passes, the delta time, in which although the simulation time does not proceed, a check is made to ensure that the right-hand sides of all assignments have already been evaluated before the new value of the assignment under consideration becomes effective. Otherwise the parallel processing of instructions would not be possible.

In the case of an analogue circuit, the modified node voltage analysis is generally used, see Vlach and Singhal [410] for a good overview. This establishes differential equations for capacitances and inductances. Transistor models can include one or more parasitic capacitances. Otherwise the heart of transistor models, like diode models, is made up of a parallel circuit consisting of a resistor and a current source, the parameters of which have to be set for each new time interval. This corresponds with an arbitrary linear characteristic that can be placed as a tangent at the current working point on the nonlinear characteristic of the transistor. Voltage and current sources each correspond with constraints that are formulated in algebraic equations. Resistors are also expressed in algebraic equations. Overall a differential-algebraic equation system is established that is also known as DAE

---

[2] See Section 6.2.

(differential-algebraic equation set). The number of equations depends upon the circuit and is very high, typically significantly above the number of degrees of freedom. The resulting system matrices are however only sparsely occupied.

For multibody mechanics, the equations of motion are normally derived by means of the application of a classical principle, e.g. that of Lagrange or D'Alembert. When drawing up the equations it is possible to choose between two extremes. In one case the generalised coordinates, which fully describe the state of a system and which can also be regarded as degrees of freedom, are first determined. For n generalised coordinates (at least for holonomous systems) n equations can be drawn up. The constraints fall away, leaving a system of ordinary differential equations. However, these may turn out to be very complex. Alternatively, it is possible — as in electronics — to permit more unknowns and thereby obtain a system of differential equations for the motion of bodies and algebraic equations for the constraints, which may, for example, be caused by joints. This establishes a system of DAEs, which can be solved using similar methods to those used in the circuit simulation, see for example Orlandea *et al.* [304]. In both cases the number of degrees of freedom is relatively small in comparison to those in electronics. The number of objects under consideration, such as bodies, joints, springs, shock absorbers, etc. is generally significantly below one hundred. However, the numerical problems caused by transitions between static and sliding friction, mechanical impacts, three-dimensional coordinate transformations and other effects, cannot be disregarded.

In the representation of continuum mechanics by means of finite elements the number of degrees of freedom is significantly higher than those in multibody mechanics. The associated system matrices normally have a band shape, which the simulation exploits by suitably customised numerical procedures. Overall, this normally establishes a system of ordinary differential equations, the parameters of which, i.e. the inputs into the mass, damping and stiffness matrix, may however have to be recalculated at runtime.

2.4.4 Experimental modelling

Introduction

Experimental modelling consists of the development of mathematical models of dynamic systems on the basis of measured data or at least providing existing models with parameters. So neither the underlying physics nor the internal life of the system need necessarily play a role in model generation. In contrast to physical modelling there are procedures for experimental modelling in which the modelling can be wholly or partially automated.

Table model

The simplest method of incorporating measured data is by the formulation of table models that lead to a stepped or piece-wise linear characteristic. The problem with the trivial conversion of a table model is the abrupt changes or kinks that are caused by the fact that only a finite number of values are available. The difficulties are numerical in nature since numerical oscillations may occur at abrupt changes and kinks. These are caused by the fact that — as a result of feedback — different sections of the characteristic are approached alternately and this may impair or even prevent the convergence of the simulation. A possible solution is offered by procedures that smooth the characteristic, such as the Chebychev or Spline approximations.

Parameter estimation and system identification

In this connection we can differentiate between two aspects: Parameter estimation and system identification. Parameter estimation requires a model and considers the parameters that belong to it. Some parameters, such as mass or spring constants are generally accessible without parameter estimation, whereas other parameters, e.g. coefficients of friction, can often only be determined within the framework of parameter estimation. The identified parameters then ensure the best possible correspondence between simulation and measurement.

In system identification, on the other hand, a model for the system is created on this basis or selected from a group of candidates. This is generally efficient and numerically unproblematic. The quality criterion here is the degree of correspondence that can be achieved using parameter estimation. The two significant disadvantages of parameter estimation and system identification are that, firstly, a measured result must be available in advance, which means that the system can only be considered after its development and manufacture. Secondly, the results are often not transferable, or at least not in a straightforward manner, to variations of the system or of components.

There are typically four stages to a system identification, see for example, Kramer and Neculau [206] or Unbehauen [405] and Figure 2.5.

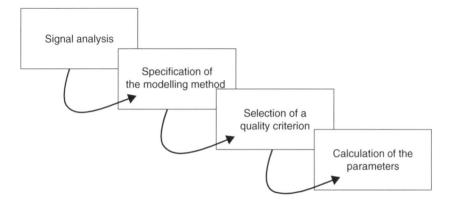

Figure 2.5 System identification sequence

The first stage of signal analysis is the establishment of a suitable test signal, which is triggered by the system. Possibilities here are step functions, rectangular pulses, triangular pulses and many more. An inspection in the frequency range facilitates investigations into whether the system to be identified is sufficiently excited over the spectrum of interest. Measurements are generally only made at discrete time points, so that a sampling interval must also be determined. Furthermore, a measurement time must be specified, the lower limit of which is determined by the point at which sufficient data is available for identification. The progressive nature of a real system imposes an upper limit on the measurement time. Then, signal processing procedures may also be used, such as averaging, root mean square calculation, or Fourier analysis, correlation analysis, and spectral analysis. So, for example, statistically dispersive noise signal components can be disposed of by averaging similar measurements which, however, multiplies the measurement time.

In stage two, determining the model approach, we can choose between prefabricated and customised structures, see for example Ljung [234]. The former may, for example, consist of canonical models in the state space and lead to a 'black-box' parameterisation, i.e. model structure and parameters have no physical significance, but rather serve merely as a vehicle for reflecting the observed behaviour. Customised equation system structures, on the other hand, are based upon a physical modelling of the system, so that the identified parameters also possess a physical significance. In any case, however, all available information about the system should be fed into this. This applies in particular to the faults that are virtually always present, which in most cases rule out an exact solution.

The identification typically rests upon minimising the discrepancy between measurement and simulated behaviour or a functional of this. Various quality criteria are used for this, one of which is selected in the third stage. Criteria are particularly frequently selected that assess a quadratic function of the measurement error.

To conclude the identification, numerical procedures are used in order to minimise the quality criteria selected in the third stage. These procedures are performed for all model structures proposed in the second stage, so that not only are the parameters in question determined in this stage, the quality of the individual structures in relation to one another are also established. This facilitates a selection of the model structure.

In the simplest case we can, as in Kramer and Neculau [206], quote the following equation for the system under investigation:

$$y_k = a \cdot x_k + n_k \tag{2.1}$$

where x_k denotes an input quantity, y_k an output quantity, n_k a disturbance variable in relation to the measurement and a is the parameter to be estimated. This relationship should be modelled on the basis of the following approach:

$$\hat{y}_k = \hat{a} \cdot x_k \tag{2.2}$$

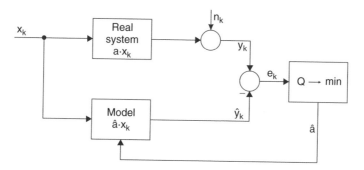

Figure 2.6 Comparison between real system and model for parameter estimation

This can also be graphically represented as shown in Figure 2.6. The aim of this is to minimise the quality function Q, so that the estimated parameter â is optimised in relation to Q.

A common approach for the quality function Q is to find an expression that is proportional to the quadratic average of the error signal e_k:

$$Q = \sum_{k=1}^{n} e_k^2 = \sum_{k=1}^{n} (y_k - \hat{y}_k)^2 = \sum_{k=1}^{n} (y_k - \hat{a} \cdot x_k)^2 \qquad (2.3)$$

where n is the number of measurements. For a compact representation the signals should henceforth be regarded in the form of n-dimensional vectors:

$$\begin{aligned}
\mathbf{x}^T &= [x_1 \ x_2 \ldots x_n] \\
\mathbf{y}^T &= [y_1 \ y_2 \ldots y_n] \\
\hat{\mathbf{y}}^T &= [\hat{y}_1 \ \hat{y}_2 \ldots \hat{y}_n] \\
\mathbf{e}^T &= [e_1 \ e_2 \ldots e_n]
\end{aligned} \qquad (2.4)$$

Thus the quality function can be described in vector notation as follows:

$$Q = \mathbf{e}^T \mathbf{e} = (\mathbf{y} - \hat{a}\mathbf{x})^T \cdot (\mathbf{y} - \hat{a}\mathbf{x}) = \mathbf{y}^T\mathbf{y} - 2\hat{a}\mathbf{y}^T\mathbf{x} + \hat{a}^2\mathbf{x}^T\mathbf{x} \qquad (2.5)$$

Now Q should be minimised in relation to â. For this to be achieved the partial derivative of Q in relation to â must become zero, i.e.:

$$\frac{\partial Q}{\partial \hat{a}} = -2\mathbf{y}^T\mathbf{x} + 2\hat{a}\mathbf{x}^T\mathbf{x} = 0 \qquad (2.6)$$

Solving this with respect to â finally gives:

$$\hat{a} = \frac{\mathbf{y}^T\mathbf{x}}{\mathbf{x}^T\mathbf{x}} \qquad (2.7)$$

Equation (2.7) is also called a regression and represents the solution for the method of least squares [206]. The inclusion of information on the interference process allows us to obtain better parameter estimates, as is the case in the weighted method of least squares.

2.5   Model Verification and Validation

2.5.1   Introduction

As defined in Section 2.1, model verification answers the question of whether the implementable model reflects the conceptual model within the specified boundaries of accuracy, whereas the purpose of model validation is to show whether the implementable model is suitable for fulfilling the envisaged task within its field of application. In what follows the most important methods in this field will be introduced. These originate from a very wide range of fields of application, some of which lie outside the field of engineering sciences. They are, however, general enough to be used in a technical context. Good overviews of the underlying literature can be found in Kleijnen [193], Cobelli *et al.* [72] and Murray-Smith [288], [289].

2.5.2   Model verification

Verification on the basis of the implementation methodology

The most direct form of verification takes place as early as the implementation stage and aims to ensure that, where possible, the errors to be identified by verification do not occur at all. This requires intervention into the methodology of model implementation. In this context, the same boundary conditions often apply as those for the development of software since, in this field too, a formal description based upon syntax and semantics is used for the formulation of a given technical content. Accordingly, most of the mechanisms that are used for software development also come into play here in order to avoid implementation errors. A few key words here, see Kleijnen [193], are: Modular modelling, object-oriented modelling or the 'chief modeller' principle, in which the actual implementation is as far as possible performed by a single person, whilst the other colleagues of the 'chief modeller' relieve him of all other tasks. In addition, there is the modular testing of submodels, so that modelling errors are recognised as early as possible and at lower levels. A further important aspect of verification lies in the correct definition of the scope of the model and in the ongoing checking to ensure that this scope is adhered to. Extrapolations beyond the guaranteed range should generally be treated with extreme caution.

Plausibility tests

Plausibility tests can also make a contribution to verification (and validation), see also Kramer and Neculau [206]. This is particularly true if they can be performed by means of simple manual calculations. They are based upon analytical considerations or the results of an initial simulation. The following criteria could possibly be drawn upon for plausibility tests:

Causality The cause should precede effect in reality and in the model. Any deviation from this principle indicates serious deficits in the model.

Balance principles The principles of the conservation of energy and matter apply not only to the physical reality, but also for the model itself.

Current/voltage laws Currents, forces and moments at a point add up to zero. Voltages and velocities add up to zero in a closed loop. These relationships apply for any electronic or mechanical system with concentrated parameters.

Value range State and output variables and parameters are normally associated with an applicable range of values. Although this is not necessarily precisely defined, unrealistic values can be recognised very quickly. For example; areas, volumes, energies and entropies can never be negative.

Consistency of units Model equations are generally formulated without units. Nevertheless, it is often worthwhile using the consistency of units as a criterion for verification.

Verification on the basis of alternative models

There are often several methods or tools available for modelling and subsequent simulation. If two approaches are independent of each other in terms of methodology and realisation, then they can be used for mutual verification. This arises because the probability of different errors producing the same effects falls, as the number of independent simulation experiments rises. Still simpler is the case where an approach has already been verified. In this case verification is established directly by means of a sufficient number of experiments, and a comparison between the model that has already been verified and the model to be verified. We see from this that absolute verification remains limited to a very small number of fields of application. In all other cases it is much more a case of deciding how many experiments must be performed before we are prepared to regard a model as having been verified. In this context, moreover, the required degree of correspondence, and consequently the accuracy of the model, has to be defined in advance.

Let us now illustrate this verification procedure on the basis of a few examples. We can use a logic simulator for the simulation of digital circuits, or — when considering the underlying transistor circuit — a circuit simulator can also be used. In principle, both simulators should deliver the same results, with the circuit simulator giving greater accuracy at a higher cost as a result of its analogue consideration method.

For simplified applications it is often possible to put forward analytical solutions that can be used for verification purposes. An example of this is the mechanical deformation of a rectangular or round plate under load, which can be calculated very simply in the form of an analytical equation. The resulting elastic line provides a starting point for the verification of the implementation of finite, mechanical elements.

Verification based upon visual inspection and animation

Another important verification method is the visual inspection ('eyeballing', see Kleijnen [193]) of the sequence of a simulation using a debugger or comparable tool. Simulators for hardware description languages often offer the use of such tools, which permit the representation of sequential modelling code as it is processed. Other forms of visualisation are represented by marks in Petri nets or the current state in state diagrams. However, visualisation can be used not only for the evaluation of the simulation process, but also for the representation of the simulation results. This is also vital because the resulting columns of figures are generally unsuitable for providing an overview of the system behaviour. The simplest and most widespread form is the x/y diagram, the x-axis of which is often time. In the field of electronic circuits this is usually sufficient. However, for the evaluation of mechanical behaviour, this is often not the case. In such cases animation procedures facilitate a better evaluation of the simulation results and thus better verification. It is self-evident that the animation, like any other tool to aid understanding of a model, also makes a contribution to validation, but this is not the subject of the present chapter.

Verification of the runtime behaviour

Occasionally tools are used that identify those parts of a model that contribute significantly to the running time. The classic approach to this is to determine the instruction currently being processed at regular intervals. This sampling allows us to obtain statistical information on the frequency of the execution of instructions and modules. This is entirely sufficient for the given purpose, but does not overload the running time of the programme under investigation. The information extracted can be used to selectively accelerate a model, which is of decisive importance particularly for more complex models which already have considerable running times.

Formal verification

Formal verification will be considered here from the point of view of formal methods for the verification of digital circuits originating from microelectronics. Since the design of digital circuits is increasingly based upon modelling in hardware description languages, we can no longer differentiate the verification methods for the designs from the verification of the corresponding models. Now if the design and simulation models are exactly the same, there is no need for verification. Occasionally, however, models have also been specially prepared for the simulation, which may be necessary for performance reasons. In this case it may be useful to perform a formal verification. This can be divided into two main fields: 'equivalence checking' and 'model checking'.

In the first case we are concerned with the functional comparison of a description with a reference description. One example could be the comparison between a gate net list and a reference model on register-transfer level, which has been intensively simulated during the design process. This largely corresponds with the verification based upon alternative models. However, in this case simulation results are not compared, as is the case for the alternative model verification. Instead formal, mathematical methods are used to find proof of equivalence.

'Model checking', on the other hand, is concerned with using mathematical methods to verify certain predictions about a circuit. So, for example, for a traffic light circuit you could exclude the possibility of all sides showing a green light [211]. This is based upon the automatic construction of a formalised proof for the prediction in question. A similar principle is followed by Damm *et al.* in [77] for the formal verification of state diagrams of automotive systems. 'Model checking' can also be used for the validation of a model.

2.5.3 Model validation

Introduction

The validity of a model is always partially dependent upon the desired applications. This is clearly illustrated by the validation criteria listed below, see also Murray-Smith [288]:

Empirical validity   Correspondence between measurements and simulations.

Theoretical validity   Consistency of a model with accepted theories.

Pragmatic validity   Capability of the model to fulfil the desired purpose, e.g. as part of a regulator.

Heuristic validity   Potential for testing hypotheses, for the explanation of phenomena and for the discovery of relationships.

These different validation requirements are the reason for the development of a whole range of validation strategies. In addition to the methods presented in the following sections there are also a few basic strategies that improve the degree to which models can be validated. In general, simpler models are easier to handle, and thus also easier to validate. In some cases it is also a good idea to take the model apart and then validate only the components and their connection together. Finally, it is occasionally worthwhile to selectively improve the quantity and quality of the measured data from the real system, which can, for example, be achieved by a design of the experiment layout that is tailored to the problem.

Direct validation based upon measured data

Validation should ensure the correspondence between the executable model and reality. To achieve this it is necessary to take measurements on real systems in order to compare these with the results of a simulation. Models are often used to obtain predictions about the future behaviour of a system. If this model is predictively valid, it follows that the predictions are correct in relation to reality. However, the reverse is not necessarily true! It is quite possible for faulty models to produce correct predictions by coincidence. So we cannot say that a model is valid on the basis of simulation experiments, but at best that the model is not valid if false predictions are made. In principle a greater number of simulation experiments does not change the situation. Only the probability that the model is predictively valid increases with the number of experiments.

The possibility of performing experiments in reality and recording their results by measurement is limited. Correspondingly, the available data tends to be scarce in some cases. As a result of the lack of support points, this can cause difficulties in validation. But the opposite case can also lead to problems. If plenty of measurement data is available, a great deal of effort is occasionally necessary to extract the relevant content from the data.

An initial clue is provided by the visual comparison of measured data and simulation results in order to ensure that the input data of the model is represented as precisely as possible in the simulation. Furthermore, a whole range of measured variables can be used to check the correspondence between measured data and simulation results. So it is possible, as demonstrated by Murray-Smith in [289], to define various Q functions for the time-discrete case, which represents a degree of correspondence between the measured response z_i and the result of the simulation y_i. The following formula shows the first possibility:

$$Q_1 = \sum_{i=1}^{n} (y_i - z_i) \cdot w_i \cdot (y_i - z_i) \tag{2.8}$$

where w_i denotes weight. This formula can also be viewed as a weighted variant of equation (2.5). Another possibility is to use Q_2 to define a normalised degree of inequality:

$$Q_2 = \sqrt{\sum_{i=1}^{n} (y_i - z_i)^2} \bigg/ \left(\sqrt{\sum_{i=1}^{n} y_i^2} + \sqrt{\sum_{i=1}^{n} z_i^2} \right) \tag{2.9}$$

The values of Q_2 lie between zero and one, with values close to one indicating a high level of inequality and values close to zero indicating a high level of equality between measurement and simulation. A further approach is recreated in the target functions of simulated annealing and genetic algorithms:

$$Q_3 = \frac{1}{1 + \dfrac{1}{n} \sum_{i=1}^{n} (y_i - z_i)^2} \tag{2.10}$$

In this case values close to one indicate a good correspondence and lower values indicate a correspondingly poorer agreement.

Although these measures achieve a significantly better quantification of the correspondence between measurement and simulation than the visual comparison, unresolved problems remain. For example, in some cases it is worthwhile to derive the individual values and draw upon general properties for comparison. One possibility is to make a comparison over the frequency range instead of over time, see Murray-Smith [289].

Validation based upon a system identification

One significant criterion for the validation of a model is how well or badly it can be identified, see previous section on parameter estimation and system identification. Cobelli *et al.* [72] classify the validation methods as identifiable and nonidentifiable models, whereby the former is described as the simpler and the latter as the more complex model. The applications considered stem from the field of physiology and medicine.

If a model is clearly identifiable then the procedure of parameter estimation can be used to validate a predetermined model structure. In the first step the parameters of the model are identified to minimise the difference between measured and simulated data. Then the following information can be obtained about the validity of the model structure:

A high standard deviation of the estimated parameters in the identification for various sets of measured data can indicate an invalid model, but it can also indicate non-negligible measurement errors.

Systematic deficits in the approximation of the measured values by the simulation indicate that the structure of the model does not correctly reflect reality.

Conversely, differences between identified and any known, nominal parameters can be evaluated. This is particularly interesting if the variance of the individual parameter estimates is known.

Furthermore, it is also possible to subject the identified parameters to a plausibility analysis. In this connection, all available information on the system should be used to discover inconsistencies in the identified parameters.

Most procedures and tools for system identification are only suitable for linear models. Furthermore, various aspects of even nonlinear models can be considered if a linearisation is performed.

Validation based upon the 'model distortion' approach

The 'model distortion' approach, see Butterfield [54] and Cameron [58], is similar to validation by identification. The main idea behind this is to calculate the 'distortion' of parameters necessary to obtain simulation results that precisely correspond with the measurements for every point in time. The gap between nominal parameters and the newly determined parameters, which alters from one moment to the next, becomes a measure for the quality of the model. In particular, it is possible to investigate whether these new parameters lie within an accepted variation of the nominal parameters. Once again, measuring precision is a problem in this approach, and this can significantly limit the value of the possible predictions. The 'model distortion' approach was originally used for the validation of models for heavy water reactors.

Validation based upon a sensitivity analysis

It is not generally possible to precisely determine the value of the parameters of a simulation model. However, it is almost always possible to define intervals within which the value of a parameter always lies. The value of a model is questionable if the variation of a parameter within the interval leads to significant variations in the simulation results. This is generally because parameters enter the model behaviour in nonlinear form. In such cases, sensitivity analysis can supply important indications of validity problems, see Kleijnen [193]. In the simplest case, the sensitivity S is determined using the perturbation method for a property of the circuit F and a parameter P, by varying the parameter by ΔP and evaluating the change in the circuit value ΔF:

$$S = \frac{\partial F}{\partial P} \approx \frac{\Delta F}{\Delta P} \tag{2.11}$$

It is often worthwhile to standardise the sensitivity in this connection:

$$S = \frac{\partial F/F}{\partial P/P} \approx \frac{P \cdot \Delta F}{F \cdot \Delta P} \tag{2.12}$$

However, this can lead to problems if F or P are close or equal to zero.

Validation based upon a Monte-Carlo simulation

The sensitivity analysis described in the previous section allows us to investigate the effects of a parameter or possibly to set the individual sensitivities of several parameters off against each other. Now the parameters and their variations are not independent of each other with regard to their effect upon the events of the simulation. On the other hand, for reasons related to the running time it is not possible to itemise all combinations of parameter variations and subject each to a sensitivity analysis. Nevertheless, in order to do justice to these cross-sensitivities to some degree we can predetermine intervals and statistical distributions for the 'suspect' parameters and run a large number of simulations, each with statistically dispersive parameters. However, we cannot prove the validity of the simulation in this manner, we can only say that the check has failed, or has not failed, after a certain number of experiments. In the former case the matter is clear, in the second the risk of the failure of validity has in any case been reduced. For this reason, this method is also called risk analysis by Kleijnen [193]. The methodology described is already built into many circuit simulators. It is generally not used for the validation of models, but for the evaluation of the yield of fabricated circuits taking into consideration the component tolerances.

Validation based upon model hierarchy

This method aims to achieve the validation of a model based upon the validation of its components, whereby the interconnection of the components occurs directly within the model and thus is noncritical in relation to validation.

A simple example of this is the validation of the model of a circuit, where this is described in the form of a net list of components such as transistors, diodes, etc. If we assume that the net list represents the actual connection structure of the circuit, then the validation of the circuit model is transformed into the validation of the component model. If only a few component types are used, which can be individually modified by parameterisation to give the desired component, then the validation of all circuits created from these components requires only one validation of the component model. Thus the validation of circuit models can in principle be considered as having been solved. The only further point of interest is the consideration of macromodels for circuit blocks such as operational amplifiers, which offer advantages in terms of simulation speed due to more abstract modelling.

A similar approach is also followed in the object-oriented modelling of multibody systems or in the creation of block-oriented models for control engineering, although the diversity of basic models is significantly greater in these cases. An example of this is the 'open loop' simulation method described by Gray and Murray-Smith [123], in which a system model is broken down into component models, which are each individually simulated with real measured data at the inputs. An example application for this is the rotor dynamics of a helicopter.

Validation based upon inverse models

In [44] Bradley *et al.* consider the modelling of a helicopter. To validate the developed model, flight trials are performed in which the pilot has to perform a predetermined manoeuvre. His control inputs are used as the stimuli for the simulation. A validation of the model cannot be achieved for certain manoeuvres because the pilot and helicopter form a control loop in which even the smallest deviations quickly accumulate to form large discrepancies between reality and simulation. His measured control movements are correct only for reality. In order to achieve a validation nevertheless, Bradley *et al.* propose to consider also the inverse of the simulation. In this case the desired flight movements are predetermined. An inverse model in the form of an ideal pilot calculates the necessary control of the helicopter. This avoids the accumulation of faults described above. Thus the validity of the helicopter model is demonstrated on the basis of outputs supplied from the inputs generated using the inverse model. The criteria from the previous section on direct validation based upon measured data, can again be applied here.

2.6 Model Simplification

In some cases the precision of some (sub)models is greater than is necessary for the purposes of the simulation. This is not critical as long as the efficiency of the simulation is not a problem. However, if the simulation times become too great then it makes sense to consider the simplification of models, see for example Kortüm and Troch [203] or Zeigler [435]. According to Zeigler the following strategies can be drawn upon to achieve the simplification of a basic model:

- Omission of components, variables and/or interaction rules.

- Replacement of deterministic descriptions by stochastic descriptions.

- Coarsening the value range of variables.

- Grouping of components into blocks and combining the associated variables.

The first method assumes that not all factors are equally important for the determination of the behaviour of a model. Typically, these factors are classified as first and second-order effects. The behaviour of a model usually depends primarily upon a few first-order effects, whilst the second-order effects, although numerous, can generally be neglected without significantly detracting from the validity of the resulting model. Here too the principle applies that the validity of a model is always established from the point of view of the application. A further difficulty is that the omission of components, variables or interaction rules can have side effects for other parts of the model. For example, an eliminated variable may leave behind gaps in various interaction rules, which each need to be carefully closed. This process is not trivial.

The second principle is based upon the observation that in many cases a stochastic formulation is significantly more simple to create than a complete deterministic description. Thus, in the investigation of the performance of a computer, for example, a proportionately weighted mix of instructions is used, instead of considering individual programmes and their sequence.

The third point recommends the coarsening of the value range of variables, such as occurs in electronics in the transition from an analogue to a digital consideration. In this approach, the variables, and of course also the components and interaction rules, are initially retained. But one value now covers a whole value interval in the original model, the individual value of which can no longer be activated. This may lead to changes in the formulation of interaction rules.

Finally, the fourth principle is based upon the combination of components and variables. For example, the distortion of a capacitive pressure element in space can be determined by a large number of positional variables. From an electrical point of view, however, only the resulting capacitance of the structure is of interest and not the strain. The capacitance, on the other hand, represents a single numeric value which is, however, partially determined by the mechanical strain.

All these methods thus serve to obtain a simulatable description from the more theoretical basis of a conceptual model without, in the process, losing the validity of the application cases of interest.

2.7 Simulators and Simulation

2.7.1 Introduction

The models introduced in the previous sections can be automatically evaluated in numerous ways. This is called simulation.[3] Before electronics came into being, attempts were made to construct mechanical equipment that displayed the same relationships between the variables as was the case in the model. Worth mentioning in this context is, for example, the tide prediction device (1879) by Lord Kelvin or the mechanical differential analyzer (1930) by Vannevar Busch. After the second world war the development of electronics resulted in the analogue computer, which was successfully implemented in the aircraft industry, for example. The field of simulation gained new impetus with the introduction of the digital computer, which brought the advantage that adaptation to a new simulation problem did not require changes to the hardware, but only different software. Today we differentiate between a whole range of simulator classes in the field of application of mechatronics and micromechatronics, the most important of which are listed in Table 2.1.

2.7.2 Circuit simulation

A circuit simulation considers networks of components such as transistors, diodes, resistors, capacitors, coils, etc. The variables that are of interest here are generally voltages and currents. These are represented in continuous form. Nonlinear, differential-algebraic equation systems have to be solved, which arise as a result

---

[3] The word simulation is derived from the Latin verb simulare, which means to feign.

Table 2.1   Classes of simulators for mechatronic systems

Simulator class	Elements considered
Circuit simulator	Circuits made up of electronic components, e.g. transistors, resistors, capacitors, coils, etc. and analogue hardware description languages
Logic simulator	Logic gates, e.g. AND, OR, NAND, NOR, XOR, etc., plus digital hardware description languages
Block diagram simulator	Block diagram of control technology
Multibody simulator	Bodies with mass and inertia moments, joints, springs, dampers, actuators, sensors, etc.
FE simulator	Finite elements for the description of a mechanical continuum
Software simulator	Programs in assembler and in higher programming languages

of the structure of the circuit. The most important procedure in this context is the modified nodal analysis. As the name suggests, nodal analysis considers the node voltages as an unknown. The important point here is that the number of node voltages, and thus the number of equations, is typically significantly higher than the number of degrees of freedom.

The process for drawing up the equation system begins with the generation of the equations for each branch, e.g. for each component in the circuit. Then there is the adjacency matrix, which describes the connection structure of the circuit and thus the relationship between branch and node voltages. Furthermore, capacitances and inductances have to be taken into account in the form of a numeric integration. The procedures of Gear, trapezoidal or backward Euler integration are often used here. Finally, the nonlinear components, such as transistors and diodes must be taken into account by bringing about a linearisation at the working point typically using the Newton–Raphson procedure.

2.7.3 Logic simulation

It is often not possible to perform circuit simulation for larger circuits[4] as a result of the associated cost. If we still want to analyse these circuits by simulation, sacrifices must be made in accuracy. For digital circuits it is generally possible to use a number of logical values e.g. (0,1,X,Z) instead of the continuous potentials used previously. Here X represents an unknown and Z a high-ohmic state. Nonideal signal changes are represented in the digital world by signal transitions, which are, however, subject to a time delay. Furthermore, time is no longer continuous, but is considered as discrete or event-oriented depending upon the simulator. Only in the latter case is a precise consideration of gate and block delays possible,

---

[4] >10 000 components.

which means that virtually all logic simulators on the market currently work in an event-oriented manner.

The two main strategies for simulation are the compiled and the interpreted simulation. In the compiled simulation, the circuit is translated prior to simulation into a programme, the processing of which brings about the simulation. In the simplest case the circuit contains a few gates and no feedback loops. In this case an instruction can be provided for every gate, which applies the function of the gate at the gate's inputs and stores the result as a variable, which represents the output signal of the gate. These instructions are sorted topologically, so that all values are already calculated if they are needed in a calculation. If the circuit contains back-couplings this principle can no longer be maintained. In this case we work with two sets of variables — one for the values drawn into the calculation and one for the newly calculated values. Then we iterate until no more changes occur in the circuit.

By contrast, in the interpreted simulation, information is available for every gate about which other gates are connected to it. The idea is to not recalculate all gates afresh for each step, but only to calculate those for which the logical value of the input has changed. For all other gates nothing has changed. We start with the inputs to the circuit and evaluate the gates connected to it. In this manner, future events are generated at the gate outputs in question, which are stored in a chronologically sorted list. Based upon this list the next event in chronological terms can be determined and the associated gate calculated, which often triggers further events. In addition, further events may occur at the circuit inputs. The simulation ends when the event list is empty or a predetermined time period has passed.

2.7.4 Multibody simulation

In this context we can differentiate between two main types of mechanical simulators, see Leister and Schielen [233]. Firstly, there are the simulators that formulate the mechanics as a symbolic equation system, which can then be processed using numerical standard solution procedures. The other option is to consider the mechanics as a linear system with mass, damping and stiffness matrix. In this case, the individual coefficients of the matrices have to be determined afresh for every time increment. Both approaches have their advantages and disadvantages.

Depending upon the application under consideration, the symbolic equations may explode in size, putting them beyond any simulation. On the other hand, there are occasionally numerical advantages, because, in the case of symbolic equations, the equation solver is in possession of all relevant information about the system. This is not the case for the numerical variant because the calculation of system matrices typically tends to be independent of the differential equation solver. Finally, symbolic equations can also be used in another context, for example, optimisation.

2.7.5   Block diagram simulation

A block diagram[5] describes the structure of a system of mathematical equations in graphical form. All connections between blocks are set up so that the causality of the system can be determined in advance. This facilitates the creation of a simulator for block diagrams that explicitly builds up system equations. Thus a sequence of instructions can be processed during the simulation. This is more efficient than an implicit formulation using 'genuine' equations. However, the causality must be determined in advance and must not change during the course of the simulation, which may well occur, depending upon the system.

2.7.6   Finite element simulation

Every finite element is characterised by its mass, damping and elasticity matrices. These matrices are square; the number of rows and columns corresponds with the number of degrees of freedom of the element. For small deflections the movement of the mechanics can still be considered as linear. In this case it is sufficient to establish the element matrices once at the beginning of the simulation. Otherwise, the element matrices must be calculated afresh for every time increment.

In order to determine the behaviour of the entire structure, the element matrices are now converted to system matrices. If two suitable degrees of freedom of two neighbouring finite elements are linked together, then on the system level they come together into one degree of freedom. In this manner, various degrees of freedom are dispensed with on the element level. The resulting system matrices have a band structure, for which certain solution procedures, e.g. the Cholesky method, are particularly suitable.

2.7.7   Software simulation

The most obvious form of software simulation is performing it on a computer. A debugger is generally available for this, which displays the processing of programme instructions and the current variable value and outputs. The timing of processing naturally varies according to the computer used, so that at this level only functional investigations tend to be performed.

Furthermore, there are also so-called command set simulators, which consider the software processing for a certain processor at assembler level. Timing can be determined on the basis of the timing cycles that have elapsed. This is only the case, however, if access to external resources, e.g. to a hard disk, can also be precisely specified in the timing, which is rarely the case. For embedded processors such resources are often not available, which means that in this case precise values can often be obtained for the timing.

---

[5] See also Section 3.4.2.

2.8 Summary

This chapter has presented a cross-section of the methods for the modelling and simulation of electronics and mechanics that are currently prevalent. It has listed categories and fields of application of models. It has also taken into consideration methods for the verification, validation and simplification of models. This forms the basis for the consideration of electro-mechanical systems in the next chapter.

3

Modelling and Simulation of Mixed Systems

3.1 Introduction

The majority of technical systems are mixed; i.e. they incorporate components from various fields, such as electronics, mechanics, software and other domains. This raises significant design problems because hitherto design methodologies and the associated design tools have usually been developed for a single field only. This means that the overall function of the system cannot be investigated until the prototype construction phase. However, by the time this stage is reached, changes to the design have already become very expensive and time-consuming. The consideration of virtual prototypes, which allow virtual experiments to be performed on a computer by simulation, offers an elegant solution to the problem described above.

This chapter introduces the consideration of the simulation of mixed systems by describing common ground and differences between electronics and mechanics in Section 3.2. This lays the foundation for the modelling and simulation of electromechanical systems.

This chapter also describes various approaches to the modelling of mechatronic and micromechatronic systems. One possibility is to transfer mechanical models into the form of electronic models (and vice-versa). This permits the consideration of the mechanics in a electronics simulator (and vice-versa), see Section 3.3. Thus half of the modelling problem can take place using standard methods. On the other hand, this raises the question of how to formulate electronics within the modelling world of mechanics (and vice versa). For the transformation of mechanics into a circuit simulator the use of hardware description languages is the method of choice. This approach — the main theme of this work — will be discussed in detail in Chapters 4–6. Before hardware description languages became widely accepted in recent years, equivalent circuits were often developed to describe the behaviour of mechanical components. Relatively few attempts were made to consider electronics along with mechanics in a mechanical simulator. Although some mechanical simulators permit the inclusion of simple components such as capacitors, resistors

Mechatronic Systems   Georg Pelz
© 2003 John Wiley & Sons, Ltd   ISBN: 0-470-84979-7

or inductances, the consideration of active components or entire circuits has hitherto only been realised in experiments. One possible reason for this is that when developing mechanical parts of the system it is often sufficient to describe the electronics in abstract form using controller equations and thereby to avoid the circuit level.

There are also some approaches that attempt to model the entire electromechanical system as a unit without any preference for electronics or mechanics. These methods include bond graphs, block diagrams, and modelling languages such as Modelica. Despite the elegance of these description forms it is generally found that neither the electronics nor the mechanics can be modelled with the usual standard procedures, see Section 3.4.

Finally, the possibility of coupling together simulators for different domains represents a further approach to solving the problem. This could, for example, occur systematically with the aid of a simulator backplane, as is often created for pure electronics. Typical applications for this are the coupling of circuit and logic simulators or the distribution of simulations on a parallel computer or a cluster of workstations. However, simulator coupling is associated with a whole range of problems: Firstly the resulting simulator package is unwieldy, it is often difficult to operate, and licences are required for all of the individual simulators. Secondly, the problems associated with synchronisation between two very heterogeneous simulator cores are even more severe, see Section 3.5.

At this point it should be re-emphasised that this work deals with the simulation of mixed *systems*. Electro-mechanical *components* will be considered only within the context of the system.

3.2   Electronics and Mechanics

3.2.1   Introduction

The following section will investigate the common ground and differences between electronics and mechanics and the associated models. For this purpose the modelling of the two domains will be considered on the level of an abstraction, see Figure 3.1. On the lowest level we find the consideration of electrical and magnetic fields and of the mechanical continuum. In electronics such considerations are required exclusively for the design of components, e.g. transistors, and this approach is known as device simulation. In the present context, however, we are interested in systems and therefore this type of simulation can be disregarded. Above this we find circuit simulation, which considers net lists of electronic components. In digital circuits we can convert continuous voltage levels into discrete values, such as 0 and 1, thereby significantly accelerating the simulation. Using digital electronics we can build processors on which software runs, which can itself act as an abstraction level. In mechanics, on the other hand, it has hitherto

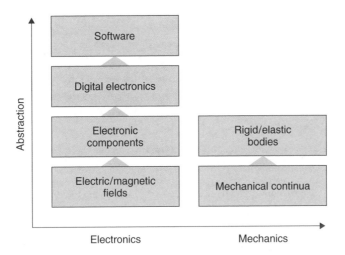

Figure 3.1 Levels of abstraction for electronic and mechanical models

only been possible to differentiate between two levels of abstraction, the continuum level and the level of multibody systems in which rigid and elastic bodies are each considered as a unit. In particular, we cannot neglect the continuum level for the consideration of systems since an electro-mechanical transformation, e.g. sensors and actuators, occasionally cannot be abstracted to the multibody level. The demonstrators from the chapter on micromechatronics are a good example of this.

3.2.2 Analogies

Analogies on the level of electronic components and mechanical bodies represent the predominant theme running through the joint consideration of electronics and mechanics. By this we mean that electronics and mechanics can be described using equations that have the same structure. This is also made clear by the fact that the equations from both mechanics and electronics can be derived from the Lagrange principle, see Maißer and Steigenberger [252] and Section 6.2.2. 'Langrange approach'. The analogies between electronics and mechanics will first be explained by means of an example, see Ogata [300]. The diagram on the left-hand side of Figure 3.2 shows a simple mass-spring-damper system.

The differential equation describing the system is as follows:

$$m\ddot{x} + b\dot{x} + kx = F \tag{3.1}$$

First we have to find out which variables can be identified as being analogous with one another. One possibility is to associate forces with currents and velocities with voltages. In order to construct an analogue circuit, let us now consider the mechanical system more closely. In this all forces act upon the mass, i.e. upon a point, and correspondingly add up to zero. In electronics this corresponds with the situation

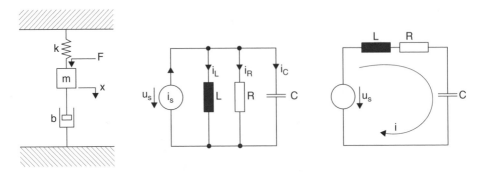

Figure 3.2   Mechanical system and two analogue circuits

in which all currents of analogue components meet at a node and also add up to zero there. Thus the circuit shown at the centre of Figure 3.2 represents an analogy with the mechanical system. Using Kirchhoff's current law the following is true:

$$i_L + i_R + i_C = i_s \tag{3.2}$$

where

$$i_L = \frac{1}{L}\int u_s \, dt, \qquad i_R = \frac{u_s}{R}, \qquad i_C = C\dot{u}_s \tag{3.3}$$

So equation (3.2) becomes:

$$\frac{1}{L}\int u_s \, dt + \frac{u_s}{R} + C\dot{u}_s = i_s \tag{3.4}$$

The magnetic flux ψ has the following relationship to the voltage u_s:

$$\dot{\psi} = u_s \tag{3.5}$$

Since voltage u_s is analogous to velocity, ψ, as an integral of the voltage, represents deflection. Thus equation (3.4) can be formulated as follows:

$$C\ddot{\psi} + \frac{1}{R}\dot{\psi} + \frac{1}{L}\psi = i_s \tag{3.6}$$

The structure of this equation exactly corresponds with equation (3.1). Capacitance is linked to mass here, damping to the inverse of resistance and the spring constant to the inverse of inductance. Lastly, the current i_s of the source corresponds with the activating force F.

Alternatively, we can also associate forces with voltages and velocities with currents. In this case the voltages, as the counterpart to the currents, must add up to zero and therefore must be arranged in a loop, see the right-hand side of

Figure 3.2. The following applies:

$$L\dot{i}_s + Ri_s + \frac{1}{C}\int i_s dt = u_s \tag{3.7}$$

If we formulate the equation with the aid of charge q, it becomes:

$$L\ddot{q} + R\dot{q} + \frac{1}{C}q = u_s \tag{3.8}$$

This equation too corresponds with the structure of equation (3.1). Now, however, the inductance is linked to the mass, the resistance to the damping, and the spring constant to the inverse of capacitance. The voltage u_s of the source is associated with the activating force F here.

We can thus differentiate between two types of analogy, which differ from one another primarily in the assignment of variables and basic elements. The force–current analogy that we investigated first has the advantage that it retains the structure of the mechanical system, see Crandall *et al.* [75]. Parallel circuits remain parallel circuits, series circuits remain series circuits. Kirchhoff's current and voltage laws apply accordingly, i.e. forces/currents at a node and (relative) velocities/voltages in a loop cancel each other out. The two Kirchhoff's analogies do not apply, if — as in the second case — forces and voltages are identified as analogous. Table 3.1 shows the most important relationships for the force-current analogy.

3.2.3 Limits of the analogies

The analogies described above are based upon linear relationships. However, this circumstance often cannot be guaranteed. For example, the Stokes' friction or viscous friction has a linear relationship with velocity in a first approximation and can thus be represented as a resistance. However, this is very definitely not the case for the Coulomb friction. Here we can differentiate between two states of static and sliding friction, for which different coefficients of friction apply. Furthermore, the Coulomb friction is not dependent upon velocity but on another variable — the perpendicular force. The Newton friction of bodies moved quickly through a fluid finally depends upon a few parameters, such as the frontal area, the drag coefficient and the density of the fluid, but above all on the square of the velocity. In order to construct an analogy for the Coulomb friction we need a resistance controlled via the normal force, i.e. via the corresponding current, which switches the coefficient of friction in an event-oriented manner upon the transition from static to sliding friction and vice versa. The Newton friction of bodies moved through a fluid, on the other hand, can best be represented as a resistance with a quadratic characteristic. We have thus already dealt with a good proportion of the components normally considered in analogue electronics.

The transition from one-dimensional to three-dimensional mechanics represents the limit of the consideration of analogies. The analogies can no longer be used

Table 3.1 Analogies between analogue electronics, translational and rotational mechanics

Analogue electronics	Translational mechanics	Rotational mechanics
Current i	Force F	Torque M
Voltage u	Velocity v	Angular velocity ω
Coil $u(t) = L \cdot \dfrac{d}{dt} i(t)$	Spring $v(t) = \dfrac{1}{k} \cdot \dfrac{d}{dt} F(t)$	Torsion spring $\omega(t) = \dfrac{1}{k} \cdot \dfrac{d}{dt} M(t)$
Capacitor $i(t) = C \dfrac{d}{dt} u(t)$	Inertia $F(t) = m \dfrac{d}{dt} v(t)$	Rotational inertia $M(t) = J \dfrac{d}{dt} \omega(t)$
Resistor $i(t) = \dfrac{1}{R} \cdot u(t)$	Damping $F(t) = b \cdot v(t)$	Rotational damping $M(t) = b \cdot \omega(t)$
Electrical power dissipation at resistor $P(t) = u(t) \cdot i(t)$	Mechanical power dissipation due to damping $P(t) = v(t) \cdot F(t)$	Mechanical power dissipation due to damping $P(t) = \omega(t) \cdot M(t)$
Magnetic energy $T(t) = \frac{1}{2} L i^2(t)$	Elastic energy $T(t) = \dfrac{1}{2} \cdot \dfrac{1}{k} F^2(t)$	Elastic energy $T(t) = \dfrac{1}{2} \cdot \dfrac{1}{k} M^2(t)$
Electrostatic energy $T(t) = \frac{1}{2} C u^2(t)$	Kinetic energy $T(t) = \frac{1}{2} m v^2(t)$	Kinetic energy $T(t) = \frac{1}{2} J \omega^2(t)$
Transformer $i_1 \cdot u_1 = i_2 \cdot u_2$ $i_1 = \alpha i_2$ $u_1 = \dfrac{1}{\alpha} u_2$	Lever $F_1 \cdot v_1 = F_2 \cdot v_2$ $F_1 = \alpha F_2$ $v_1 = \dfrac{1}{\alpha} v_2$	Gear $M_1 \cdot \omega_1 = M_2 \cdot \omega_2$ $M_1 = \alpha M_2$ $\omega_1 = \dfrac{1}{\alpha} \omega_2$
Sum of all currents at a node is zero	Sum of all forces at a point is zero	Sum of all moments at a point is zero
Sum of all voltages in a closed loop is zero	Sum of all relative velocities in a closed loop is zero	Sum of all relative angular velocities in a closed loop is zero

in this case, see Crandall *et al.* [75]. This becomes clear intuitively if we look at the example of a robotic arm. In the calculation of kinematics and dynamics, three-dimensional translational movements and triaxial rotational movements are calculated independently of one another. There is no parallel to this in electronics. Furthermore, analogies in the sense described are defined exclusively for the consideration of concentrated components and continuous quantities. Continuum mechanics, digital electronics and software thus remain outside their scope and must be considered separately.

3.2.4 Differences between electronics and mechanics

In what follows the primary differences between electronics and mechanics will once again be briefly summarized, see also Cellier [62].

With the exception of high-frequency circuits, electronics can be considered exclusively in topographic form in the simulation. The precise geometry is unimportant or can be considered using simple parameters. This is not the case in mechanics, where three components of translation and three components of rotation have to be taken into account.

Furthermore, translation and rotation cannot be considered independently of one another, as illustrated by gyroscopic forces such as the Coriolis force.

A whole range of reference systems are relevant to the description of position, movement and acceleration. We have the inertial system and various fixed body reference systems, the origins of which may lie approximately at the centre of gravity or at the coupling points. In electronics there is only a reference voltage (ground) as the 'inertial system', and voltage or current arrows as fixed-component 'reference systems'.

In electronics, and in particular in microelectronics, we sometimes have some tens of millions of components. In mechanics, at most, a few tens to a few hundreds of basic elements, e.g. rigid bodies, joints, springs, etc. have to be taken into account.

The movements of mechanical bodies are typically subject to a whole range of limitations. Mechanical stops are one example. Springs can only be extended up to a certain degree. Elastic bodies deform under the effects of force. Similar effects can also be found in electronics but they are far less prominent than is the case in mechanics.

3.3 Model Transformation

3.3.1 Introduction

We can now specify a class of simulators and use this as the basis for the description of models in the other domains. In principle, the basic simulator should be sought out on the basis of the focal point of the desired investigation. In what follows we will describe approaches based upon circuit simulators, logic or Petri net simulators, multibody simulators, and finite-element simulators.

3.3.2 Circuit simulation

Introduction

In a circuit simulator the formulation of transformed models classically takes place in a hardware description language. This approach is the main theme of the present work and will be described comprehensively in the following chapters. Alternatively, it is also possible to draw up equivalent circuit diagrams for mechanical components. We can initially differentiate between two possibilities here. Firstly, we can use the analogies introduced in Section 3.2.2 to associate electronic components with basic mechanical elements. The other option is to model not the mechanics itself, but rather the differential equations that describe the mechanics.

Analogy approach

In order to consider the analogies we must first refer to Section 3.2.2. The force/current analogy is normally used. In addition to the basic elements, other mechanical phenomena such as Coulomb friction have to be considered. These require behavioural modelling based upon sources that can be controlled by arbitrary mathematical functions. Such voltage and current sources are available in PSpice, for example. This represents a rudimentary form of modelling in a hardware description language.

Yli-Pietilä *et al.* [431] use this method to investigate mechatronic systems such as a linear drive. They model a DC motor with an electronic control system and a mechanical load. The same approach is further elaborated by Scholliers and Yli-Pietilä in [369] and applied to other examples, such as a double pendulum. In [368] Scholliers and Yli-Pietilä introduce a whole library of such models, which expand the field of application of a circuit simulator such as Spice in the direction of mechatronics.

Examples for the use of equivalent circuit diagrams in micromechatronics are supplied, for example, by Antón *et al.* [13] (pressure sensor elements), Garverick and Mehregany [111] (micromotors), or Lo *et al.* [236] (resonators).

Modelling of differential equations using equivalent circuits

As an alternative to the analogy approach described above we can also find an equivalent circuit for the underlying system of equations. In principle, this procedure is similar to the construction of a rudimentary analogue computer from electronic components. In this context we can differentiate between explicit and implicit methods, see Bielefeld *et al.* [31]. In the explicit version the values of the state variables are represented as voltages in the network. In this, the highest time derivative of each state variable is set depending upon lower derivatives and other state variables using a controlled voltage source. In addition, there are integrators, see the left-hand side of Figure 3.3, which again provide the low derivatives in the form of voltages, see Herbert [139]. As an alternative to this, the implicit method, see Paap *et al.* [312], solves a set of n equations in the form:

$$f(\mathbf{x}, \dot{\mathbf{x}}, t) = 0 \qquad (3.9)$$

where $\mathbf{x}$ represents a vector of n unknowns. As in Herbert [139] the states are represented as node voltages. Each equation is defined by a current from a voltage-controlled current source. This sets the input current of the differentiator in the

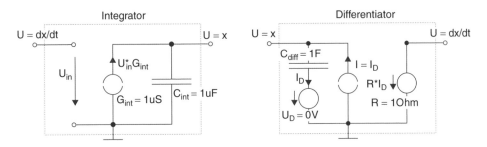

Figure 3.3 Equivalent circuits for integrator and differentiator

right-hand side of Figure 3.3 to the value $f(\mathbf{x}, \dot{\mathbf{x}}, t)$. The circuit simulator ensures that no current flows into the differentiator and thus solves the differential equation.

It is also worth mentioning that the, somewhat tiresome, process of converting a system of differential equations has been automated using the MEXEL CAE tool, see Pelz *et al.* [322]. A model transformer reads in the differential equation system, simplifies it if necessary, and then writes out a Spice net list in explicit or implicit formulation.

3.3.3 Logic/Petri net simulation

Introduction

Predicate/transition networks (Pr/T networks), see [115] represent an extension of Petri nets and are often used for the modelling of software and/or digital electronics. They permit a system description on a very abstract level in which the use of hierarchies permits particularly compact representations. The strength of Pr/T networks lies in the effective consideration of parallel processes. Brielmann *et al.* [46], [47], [48] and Kleinjohann *et al.* [199] introduce methods for describing mechanics and other physical domains, plus the associated interfaces using the resources of the Pr/T networks. Such model transformations thus provide the option of describing and simulating mixed systems in a consistent manner. The representation of the hardware description language VHDL in a coloured Petri net by Olcoz and Colom in [301] shows that Petri net simulation and logic simulation are not so very different from each other, which means that the events portrayed in the following section could well be achieved on the basis of digital hardware description languages.

Definition of Predicate/Transition nets

Pr/T nets consist of places, transitions, and directional edges between these. Places can contain identifiable markings, which represent the state of the network. If a marking is sufficiently high at the inputs of a transition and if these satisfy any additional conditions, then the transition can 'fire'. In this case the markings in question are cleared from the input places, new markings are generated at the output places and predefined actions may be performed where applicable. Such a network can be formulated in very compact form using the tools of predicate logic, e.g. in the Prolog language, see Negretto [295]. In this connection, a marking at a place conveys the information that a predicate assigned to that place is fulfilled. In order also to correctly take account of the timing of the individual components it is necessary to add in a concept of time. So in [47] two delays are assumed for a transition. One relates to the period of time for which the markings must be present at the input places before the associated transition can fire. The other describes the time that elapses between the firing of the transition and the generation of the output markings.

Modelling of discrete relationships

Let us now clarify how a Pr/T net works on the basis of a small example from [46], see Figure 3.4. On the left-hand side a piece of code is represented at the start of which some variables are initialised. There follows a loop, in the body of which various arithmetic operations are performed. The termination condition for the loop is located at the end of the loop and is based upon a comparison of two variables. In the centre and at the right-hand side of Figure 3.4 Pr/T nets are represented in different states. The variant in the middle shows the initial occupation of the markings and thus the situation after initialisation. The two calculations are located

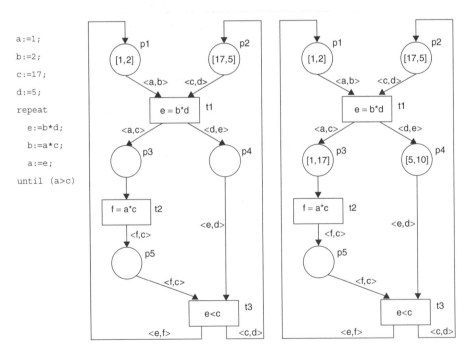

Figure 3.4   Modelling of digital behaviour using Pr/T nets

in the transitions t1 and t2. Transition t3 compares the corresponding values. If the newly calculated a is less than c, then the values are once again entered into places p1 and p2, which represent the input places of the loop. Otherwise the calculation of the loop is broken off and further transitions that are not shown can fire. The diagram on the right-hand side of Figure 3.4 shows a state in which the transition t1 has already fired for the first time. Accordingly, the new values of a and c have been entered at place p3 and the new values of d and e have been entered at place p4. Other constructs of a programming language can be depicted in the same manner.

Modelling of continuous relationships

Continuous relationships are classically modelled using differential equations that can be either linear or nonlinear. Let us now model such equations on the basis of Pr/T nets using the event-oriented modelling introduced in the previous section. A solution for linear differential equations on the basis of the Z transformation was proposed by Brielmann and Kleinjohann [46]. In what follows we will, however, predominantly consider nonlinear systems. This property can have two causes: firstly, nonlinearity can arise as a result of discontinuities; secondly, it may be caused by nonlinear functions in the system equations [47]. The first case is very simple to solve. Here only the current equation has to be activated, which can be performed by simply distinguishing between cases. The example from Figure 3.4 illustrates this state of affairs. The termination condition at the end of the loop corresponds with such a case differentiation. More difficult is the realisation of the other variant. In [47] it is proposed to put the equations together step-by-step from linear components, meaning that here too a swapping of linear components would be necessary. Furthermore, the differential equations have to be numerically integrated, which is achieved using the Euler principle:

$$\dot{x}(t) \approx \frac{x(t+h) - x(t)}{h} \tag{3.10}$$

where h is the time step of the integration. Now if the differential equation of interest is

$$\dot{x}(t) = f(x, t) \tag{3.11}$$

the integration formula is found to be

$$x(t+h) \approx h \cdot \dot{x}(t) + x(t) = h \cdot f(x, t) + x(t) \tag{3.12}$$

which, along with an additional function g(x,u,t) to determine the outputs, can be directly represented on a Pr/T net, see Figure 3.5.

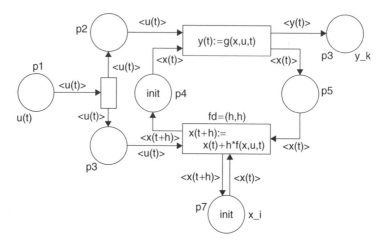

Figure 3.5   Modelling of a nonlinear differential equation using a Pr/T network

3.3.4 Multibody simulation

Introduction

In this section two approaches will be introduced: Firstly the equations of electronics will be obtained using the Lagrange principle, so that they can be seamlessly incorporated into a multibody simulator based upon the Lagrange principle. The other method is based upon object orientation, thus allowing the non-mechanical components to be modelled more or less independently of the system as a whole.

Electronic modelling using the Lagrange approach

In [253] Maißer describes a principle that uses the Lagrange approach from mechanics in order to find model equations for the electronics of a mechatronic system. In this manner the electronics can be easily incorporated into the multibody simulator, which may also be based upon the Lagrange equations. Mechanics and electronics are thus modelled using a unified approach and simulated as a whole system.

Object-oriented approach

This section introduces an approach that combines modelling on a component level with the automatic creation of a system model. As in software development this 'local' procedure is called object-orientation. Such approaches are naturally particularly well suited for describing nonmechanical parts of the system in a form that is suitable for a multibody simulator.

Kecskeméthy [185] and [186] as well as Risse et al. [346] describe a simulation environment for mechatronic systems that includes the electronics of a controller. This takes place in the form of abstract controller equations, developed using a suitable tool, e.g. MATLAB/Simulink. In this connection a class of controllers is prepared in [346] that includes continuous, proportional, discrete and mixed controllers. Simple, electronic components can also be described on the same basis. The underlying equations are added to the equations of motion of mechanics, and the equations of sensors and actuators, and are then solved as a whole.

3.3.5 Finite-element simulation

One possibility for system simulation using a FE simulator is to fuse the equation system of electronics together with the equation system of finite elements. The resulting equations include the sought-after unknowns from electronics and mechanics. The complete system can thus be processed using a standard solver.

Particularly important in this context is the work of Bedrosian [22], who expanded a finite element simulator for the calculation of electromagnetic fields so that it could process both analogue circuits and also the kinematics of rigid bodies. A significant aspect of this is to obtain a few desirable properties of FE matrices. So in contrast to the matrix for the finite elements, the system matrix would be neither positive definite nor sparse. Bedrosian therefore insists upon a separate consideration of the matrices for the individual domains, which requires a suitable iteration in order to obtain a consistent solution for the system as a whole.

3.3.6 Evaluation of the model transformation

The introduction of analogue hardware description languages has caused interest in equivalent circuits for mechanical components to fall sharply. This is primarily because a hardware description language is significantly more flexible in its formulation. This is true particularly for components for which the analogies provide no direct parallel. Furthermore, the overview is quickly lost if it is unclear what the equivalent voltages and currents represent.

In principle, the modelling of continuous relationships on an event-oriented basis — for example using digital logic or a Pr/T network — is nothing unusual. Every simulator for analogue processes that is run on a digital computer has the same fundamental problem to solve. The difference in the present case is that the basic functions of the simulation, such as the integration procedure or the automated selection of a suitable step size, have to be modelled fully by the user, which firstly can be very cumbersome and secondly presumably raises a performance problem.

When discussing the simulation of mechatronic systems in a multibody simulator it is particularly worth mentioning the elegant solution of Maißer [253], which models the electronics according to the Lagrange principle, so that the resulting equations are compatible with multibody simulation, which is also based upon the Lagrange approach. However, the lack of any significant libraries of transistor models and the fact that digital electronics and software are disregarded, are problematic.

For the variant of model transformation on the basis of a FE simulator, the field of application of a corresponding solution is only a little wider than that of the FE simulator. This is also emphasised by the fact that there is comparatively little literature in this field.

3.4 Domain-Independent Description Forms

In this section approaches will be described that cannot be classed as an expansion of the tools in a certain domain. The most important representatives here are bond graphs, block diagrams and modelling languages such as Modelica, Dymola or ACSL.

3.4.1 Bond graphs

The bond graph approach, see for example Karnopp and Rosenberg [180] or Thoma [398], fundamentally rests upon the same principles as the analogies in electronics and mechanics, see Section 3.2.2. However, there is one significant difference. In the analogies, currents were generally identified with forces/moments and voltages with velocities, so that an analogy in the form of an equivalent circuit has the same structure as the original system. This is true because according to Kirchhoff's laws, currents and forces add up to zero at a node and voltages and relative velocities add up to zero in a closed loop.

By contrast, in the bond graphs, the following classifications are made. Voltages are normally associated with forces/moments and called effort, currents are associated with velocities/angular velocities and called flow. The elements used in the bond graph approach can be divided into one, two and three-port networks. The one-port networks are the so-called C, I and R elements, which in electronics correspond with capacitors, inductors and resistors and in mechanics correspond with springs, masses and dampers, see Table 3.2. In addition there are sources for effort and flow. Transmission elements and gyrators are defined as two-port networks. The former transmit effort to effort or flow to flow in a fixed or variable relationship to one another; the latter put the effort, on the one hand, into a relationship with the flow, on the other (and vice versa). Transmission elements can thus be transformers, gears or levers for small deflections. A gyrator could for example describe a DC motor. The three-port networks finally represent serial or parallel junctions (s-junction, p-junction). The one, two and three-port networks are linked together by half arrows, so-called bonds, which each bear an effort and a flow. The direction of the arrow shows the direction of the positive power flow. The work done is found by the product of effort and flow. In addition to the half arrows of the bonds there are also connections with a full arrow, in which either the effort or the flow is neglected. These connections carry information, but no energy.

The calculation of bond graphs first of all requires the drawing up of a suitable system of equations, which is generally explicitly formulated. This means that the

Table 3.2 Assignment of magnitudes and elements in bond graphs

Bond graphs	Electronics	Mechanics, translational	Mechanics, rotational
Effort	Voltage	Force	Torque
Flow	Current	Velocity	Angular velocity
C element	Capacitor	Spring stiffness	Torsional spring stiffness
I element	Inductor	Mass inertia	Moment of inertia
R element	Resistor	Damping, translational	Damping, rotational
Transmission element	Transformer	Lever, pulley block	Gears

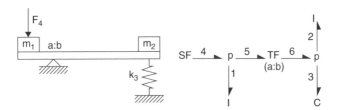

Figure 3.6 Bond graph of a simple mechanical system

equations take the form of instructions, and this fact requires a consideration of the causality of the system. Therefore, cause and effect have to be specified for each element. If we take any C, I or R element we can ask whether the effort causes the flow or vice versa. Both are possible and there are equations for both cases, which can be used in a system of equations if required. Overall, it is a question of creating continuous chains of cause–effect relationships, which can be illustrated by a suitable sequence of assignments. In the case of algebraic loops this cannot be achieved, so additional measures are necessary.

In what follows, a few examples of bond graphs will be presented. Figure 3.6 shows the bond graph of a simple mechanical system, which consists of two masses, a spring and a lever. In addition to I and C elements the bond graph contains a flow source, which represents the force F_4 and is designated SF. The transmission element TF represents the lever, which sets a ratio (a : b).

Figure 3.7 shows a simple circuit and the associated bond graphs. This again includes the flow source SF. However, this now describes a current source. The transmission element TF is also present and represents the transformer.

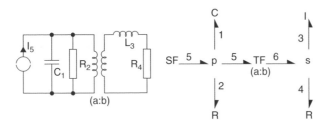

Figure 3.7 Bond graph of a simple electrical system

3.4.2 Block diagrams

Block diagrams are often used in control technology and, like bond graphs, represent a form of structural modelling, see Cellier [62]. However, this type of representation primarily shows the structure of equations, whereas the structure of the system tends to be found indirectly from the structure of the equation system.

Block diagrams include blocks and directional connections between the blocks. These connections describe signals, which are converted into other signals by the blocks. In addition there are taps and summing points, so that the important elements of block diagrams can be fully represented in Figure 3.8.

In what follows, modelling using block diagrams will be illustrated on the basis of a simple example. For this purpose we will consider the circuit on the left-hand side of Figure 3.9. This can be described on the basis of the following equations:

$$i_1 = \frac{u_1}{R_1}, i_2 = \frac{u_2}{R_2}$$

$$i_L = \frac{u_L}{L_1}, \dot{u}_C = \frac{i_C}{C_1}$$

$$u_1 = U_0 - u_C, u_2 = u_C, u_L = u_1 + u_2$$

$$i_0 = i_1 + i_L, i_C = i_1 - i_2 \tag{3.13}$$

If the above equations are translated into blocks, connections and summations, we obtain the block diagram on the right-hand side of Figure 3.9. The main problem

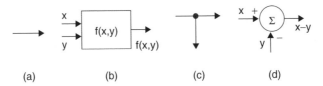

| (a) | (b) | (c) | (d) |

Figure 3.8 Basic elements of block diagrams: Connection (a), block (b), tap (c) and summer (d)

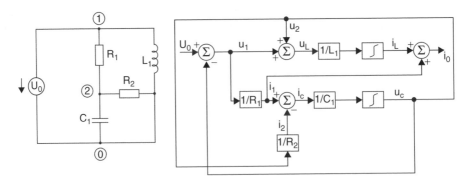

Figure 3.9 Block diagram of an electronic circuit

here is that the structure of the circuit no longer corresponds with the structure of the block diagram.

3.4.3 Modelling languages for physical systems

Languages such as ACSL [2], DSL [227], Dymola [90], [310] or Modelica [94], [272] in particular deserve a mention. All these languages support the description of physical systems. In what follows we will investigate Modelica in particular, as this language includes the most up-to-date research results and furthermore is currently being expanded to a standard, see [272] and [273]. An excellent introduction to object-oriented modelling of mixed systems in general and of Modelica in particular can be found in Otter [308].

Modelica is a language for the modelling of physical systems and was developed specifically in order to support the exchange of models and the development of libraries. Modelica does not insist upon an exclusively causal modelling, in which cause and effect of every component have to be determined even before the simulation. The description of the models can also take place in the form of genuine equations and not on the basis of assignments. Modelica supports the description of continuous systems, which can be calculated on the basis of differential-algebraic equation systems (DAE). In addition there are constructs for dealing with discontinuities, which may occur in mechanical stops, or static to sliding friction transitions. In principle it is also possible to use the discontinuities to describe event-oriented processes, e.g. transitions in a state graph or the movement of markings in a Petri net, but this possibility is limited by the underlying equation solver.

In principle, Modelica can be compared with an analogue hardware description language, see also Tiller *et al.* [400]. Both structural and behavioural modelling is possible. A particularly prominent feature of Modelica is object-orientation, which is used, for example, to declare a model — or to be specific a model class — once and instance it many times, with the option of setting certain parameters individually for each instance. Similar concepts also exist in hardware description languages, such as VHDL, with the possibilities of instancing and configuration. Modelica also offers the option of transmission between model classes, so that more complex model classes can easily be traced back to simpler ones.

To illustrate modelling in Modelica the description for an electronic circuit will be given in what follows, see Figure 3.10 and [273].

The associated Modelica model is represented in Hardware description 3.1, with key words shown in bold type. After the declaration of the `circuit` model the components along with their main parameters are declared. At this level the `equation` section specifies only the connectivity of the circuit.

```
model circuit
  Resistor R1 (R=10);
  Capacitor C (C=0.01);
  Resistor R2 (R=100);
```

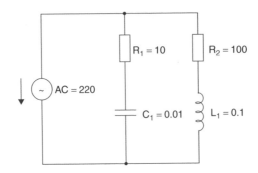

Figure 3.10 Electronic circuit as an example of a Modelica model

```
Inductor L (L=0.1);
VsourceAC AC;
Ground G;
equation
  connect (AC.p, R1.p);
  connect (R1.n, C.p );
  connect (C.n, AC.n);
  connect (R1.p, R2.p);
  connect (R2.n, L.p );
  connect (L.n, C.n);
  connect (AC.n, G.p);
end circuit;
```

Hardware description 3.1   Modelica model of the circuit from Figure 3.10

Thus the components such as resistors, capacitors, etc. remain to be described, see Hardware description 3.2. These are successively built up via the model of a pin and the model of an electrical component with two terminals. One interesting feature here is the use of inheritance in the transition from the model with two terminals to the component. Using the key word extends the roles of voltage and current and Kirchhoff's current laws are loaded into the component model and do not need to be formulated there again. An electrical component can thus be simply described by its constituent equation. In the case of the capacitor, the time derivative of voltage is designated by the function der().

```
type Voltage = Real (unit="V");
type Current = Real (unit="A");
...
connector Pin
  Voltage v;
    flow Current i;
end Pin;
...
partial model TwoPin "Parent class of the element with 2 elec.
  pins"
```

```
 Pin p, n;
 Voltage v;
 Current i;
equation
 v = p.v -n.v;
 0 = p.i + n.i;
 i = p.i
end TwoPin;
...
model Resistor "Ideal electrical resistor"
  extends TwoPin;
  parameter Real R (unit="Ohm") "Resistance"
equation
 R*i = v;
end Resistor;
model Capacitor "Ideal electrical capacitor"
  extends TwoPin;
  parameter Real C (unit="F") "Capacitance"
equation
 C* der(v) = i;
end Capacitor;
...
```

Hardware description 3.2   Model of the components from Hardware description 3.1

3.4.4 Evaluation of domain-independent description forms

From the examples shown above it is clear that bond graphs can describe both analogue electronics and mechanics (and also a range of further domains) in compact and graphic form. However, if we go beyond unidimensional mechanics and passive electronics there are significant problems to be solved. Although the modelling of transistors is also possible in principle using bond graphs, a meaningful simulation of circuits of substantial complexity remains the exclusive preserve of a dedicated circuit simulator. The same applies for three-dimensional multibody mechanics. Moreover, bond graphs are in principle limited to continuous systems, so that digital electronics and software cannot be illustrated using classical bond graphs, or at least this cannot be done efficiently. Furthermore, every element must be assigned a fixed causality prior to the simulation. This causality may alter during a simulation, for example, if an electric motor becomes a generator, so that such systems cannot be simply investigated using bond graphs. The same applies in principle for block diagrams.

Domain-independent languages, and Modelica in particular, are broadly comparable with analogue hardware description languages. However, they don't have the model basis of a circuit simulator. Furthermore, the event-oriented field is much weaker in comparison to hardware description languages in general, and VHDL-AMS in particular, so that digital electronics or software, as is demonstrated by Scherber and Müller-Schloer in [360], require the coupling of appropriate simulators to the equation solver that underlies the language.

Perhaps the most important objection against domain-independent description forms lies in the fact that it is necessary to start modelling up from scratch in every domain. Alternatively, if we build up from a circuit or multibody simulator, a large part of the system is already covered by the best available methodology.

3.5   Simulator Coupling

3.5.1   Introduction

The option of simulator coupling tackles the problem highlighted above in a straightforward manner. Appropriate simulators are already available for the various domains in the system and in the ideal case these would only have to exchange their current simulation results. The use of simulator coupling can protect investments in models and facilitate the use of the best available simulator for a field. However, simulator coupling is also associated with a whole range of problems. For example, it generally requires access to the internals of the simulators involved, which means that if commercial simulators are to be considered, the co-operation of the provider in question is required. Furthermore, the coupled simulation forms a very intricate software package, which is difficult to get to grips with. Perhaps the most important disadvantage, however, lies in the synchronisation of two normally very different simulator cores. In the coupling of analogue electronics and mechanics, differential equations are solved in both cases. However, their origin, nature and formulation are very different. Furthermore, this form of co-simulation is also associated with convergence problems, particularly in the case of a strong coupling between two analogue solvers.

3.5.2   Simulator backplane

When coupling two simulators, the principle of 'simulator backplane' represents a particularly systematic solution, see also Jorgensen and Odryna [171] or Maliniak [255]. This principle is equally suited to the coupled simulation of exclusively continuous, exclusively event-oriented or mixed systems. In principle, the simulator backplane is a standardised procedure, see Kemp [187], for the inclusion of simulators into an overall simulation, see Figure 3.11 from Zwoliński *et al.* [441]. The main task of a backplane is to undertake a partitioning of the design data before the actual simulation and to assign the individual parts of the simulators in question. The backplane also looks after the synchronisation between the linked simulators and the exchange of data. In the ideal case the backplane also has a unified user interface with the associated output tools, but this tends to be rare.

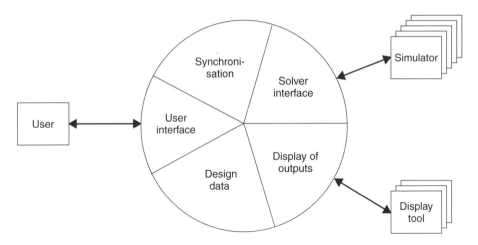

Figure 3.11   Structure of a simulator backplane

Otherwise, the corresponding settings in the individual simulators are used, which often leads to confusion. The data exchange between the backplane and the simulators can take place by means of an IPC interface.[1] This does not necessarily require that all simulators are processed on the same workstation. The load can be distributed across various computers as long as the synchronisation does not prevent this. However, the cost of communication via this comparably slow interface has to be borne. Faster simulations are generally achieved by the binding together of backplane and simulators into an overall programme. This is particularly true if a great deal of communication via the backplane is expected as a result of a strong coupling between domains in the simulated system, because in this case the addressing of a simulator from the backplane becomes a function call.

As for the general case of simulator coupling, the main problem of the backplane lies in the handling of the synchronisation between the simulator cores.[2] We can make an initial differentiation here between two classical approaches: the conservative, see Chandy and Misra [65], [66], and the optimistic, see Jefferson [168], Jefferson and Sowizral [169].

The conservative approach allows simulator A to proceed only for a time period in which it can be proven that no events sent out from other simulators are expected. These events would have to be taken into account in the simulation of A and consequently the simulation has to wait for them. In the conservative approach, simulator A is thus safe to proceed for this time period. In the extreme case, this conservative approach is called the 'lockstep' algorithm,[3] in which a fixed time

---

[1] IPC (inter process communication), communication between processes on the level of the operating system.

[2] The same problem also emerges in the parallelisation of simulations, see Fujimoto [107], where here the same simulator cores are synchronised on different processors.

[3] Other authors, such as Le Marrec *et al.* [218] or Olcoz *et al.* [302] describe the 'Lockstep' synchronisation as the specification of a global real time so that all participating simulators may each proceed their local time up to the global time.

interval is specified for all participating simulators. Particularly for systems with very different time constants this hinders an efficient processing of the simulation. In recent years a whole range of simulator couplings have been developed in the form of variations on the conservative method, e.g. Bechtold *et al.* [20], Buck *et al.* [52], Patterson [316], Sung and Ha [392], Todesco and Meng [402], Zwoliński *et al.* [441], although frequent task changes between simulators can still gives rise to performance problems in these approaches.

In the optimistic case, every simulator processes its internal events until no more activity can be determined, which in the ideal case is by far the most efficient way. Unfortunately, it may occur that another simulator generates an event for the first in this period. Then all of the first simulator's results from the moment in question must be discarded. To achieve this the simulator in question must perform a leap backwards (timewarp) and then start again at the time point in question. Depending upon the system under consideration this is associated with a high storage requirement for the saving of old states. Furthermore, depending upon the nature of the system under investigation, these timewarps can themselves become a performance problem. Normally, however, electronics simulators [402] and mechanics simulators do not provide the option of performing a timewarp, so that only the conservative approach and variations upon it remain. However, this is not necessarily the case for the co-simulation of hardware and software, see Chapter 5.

In addition to the synchronisation between two simulator cores, the question of the convergence of the solution also requires some consideration. This is particularly relevant for the coupling of two analogue cores, see Klein and Gerlach [196]. The reason for this lies in the back-coupling between the two simulator cores, which we will call A and B here. A maps its input x_A to its output y_A using a function f_A. B does the same with the function f_B. In the simplest case, the Gauss–Seidel iteration, the rule for the (k+1)th iteration step is:

$$x_B^{k+1} = y_A^k = f_A(x_A^k)$$
$$x_A^{k+1} = y_B^{k+1} = f_B(x_B^{k+1})$$

(3.14)

In this case oscillations may occur. In the worst case the iteration does not converge at all. Numerically more demanding methods, such as, for example, the Newton procedure, tend to converge better, but are not universally applicable due to the costly calculation of the required Jacobi matrices, see [196].

3.5.3 Examples of the simulator coupling

Introduction

In what follows the options and limitations of simulator coupling will be illustrated in more detail on the basis of a few examples from mechatronics and micromechatronics. This description will include the direct coupling between two simulators as well as the systematic consideration of several simulators with a backplane.

Mechatronics

In [302] Olcoz *et al.* describe the coupling of the VHDL simulator VSS with the mechanics simulator COMPAMM. Sensors and actuators are incorporated at the interface between electronics and mechanics and these are characterised by a pair of corresponding variables — one for each of the two simulator sides. The correspondence of such pairs is achieved by an interface written in C and C + +. The mechanics simulators can thus be operated using a fixed or variable time interval. In the former case the synchronisation between electronics and mechanics takes place at discrete, evenly distributed points in time that are specified by the fixed interval of the mechanics. In the latter case the mechanics simulator proceeds by a time interval and then informs the electronics simulator that it may proceed to this point. After confirmation from the electronics simulator the sequence begins again from the start.

A further approach for the coupling of simulators is mentioned by Scholliers in [367]. This approach emphasises the coupling of multibody mechanics, analogue electronics and control technology. ADAMS, PSpice and MATLAB/Simulink are the simulators used. The simulation process is centrally controlled and a fixed increment thereby specified. The application considered is a controlled drive and the mechanical load is a mechanism described in ADAMS. The actuator is a direct current motor described in the form of Spice components, whereas the PI controller exists on a purely functional level in MATLAB/Simulink.

Le Marrec *et al.* [218] describe a coupling between C routines, the VHDL simulator VSS and MATLAB/Simulink using a co-simulation bus that exchanges data between the individual simulators. The simulation can take place on two levels. Firstly, simulation can be purely functional, with electronics, mechanics, and software being investigated for the application under consideration. In the other case, the timing has to be taken into account too, necessitating a processor model in VHDL for the software. In this case the problem is merely that of the co-simulation of electronics and mechanics. The approach described is illustrated on the basis of two examples, an electronic accelerator pedal for an electric car and the control of a hydraulic suspension system for a car.

In [360] Scherber and Müller-Schloer proposed a simulator backplane that represents a mechanism for the linking of very different simulators. The approach is based upon a unified model for the heterogeneous components involved. These are termed actuators; their interfaces are called ports; every two ports can be linked by a channel. The access mechanisms are always the same. Thus the interfacing of a component and its simulation is unified without having to make limitations with regard to the nature or function of the actuators. A scheduler decides which actuators shall be executed when and for how long by means of a priority analysis. In this manner a software simulator, a simulator for finite state machines, a simulator for the Modelica language — see Section 3.4.3 – and MATLAB/Simulink were connected together.

Micromechatronics

In [121] Götz *et al.* produce a coupling between a finite element simulator and a circuit simulator. The idea consists of calculating the deformation of a pressure sensor structure using a finite element simulator and using the results to determine changes of piezo resistances. The circuit simulator is then used to determine the output of the read-out circuit. This consists primarily of a frequency modulation. Using this process the mechanical structure can be optimised very simply. However, the feedback of the electronics on the mechanics and any dynamic effects of the mechanics have not been taken into account.

A coupling between a FE simulator (ANSYS) for continuum mechanics and the circuit simulator PSpice for analogue electronics is proposed by Dötzel and Billep [86] and Klein *et al.* [198]. The applications used here are the simulation of micromirrors [86] and force sensors [198]. However, details of the coupling are not given in either case.

3.5.4 Evaluation

In simulator coupling the performance of the coupled tool is beneficial because the optimal modelling method can be selected for each field. This reduces the total modelling cost incurred, whilst the validation of the models can utilise the results within the domain in question. Further domains can be taken into account by linking in appropriate simulation tools. On the other hand, simulator coupling leads to problems in the operation of the simulator package and in handling the data flow at the interface. The simulators involved must be suitably synchronised with one another. The simulation time is typically very high due to the necessary iterations for each time interval. Finally, convergence problems may occur if there is strong coupling between the subsystems.

3.6 Summary

Various approaches to the modelling and simulation of mechatronic and micro-mechatronic systems have been considered in this chapter. We can differentiate between three groups of methods: model transformation, modelling in a domain-independent form, and simulator coupling. There are currently two options on offer that allow us to cover the whole spectrum of analogue electronics, digital electronics, software, multibody mechanics and continuum mechanics. These are simulator coupling and the universal modelling in hardware description languages, which will be described comprehensively in what follows.

4

Modelling in Hardware Description Languages

4.1 Introduction

For hardware description languages (HDL) — as for every other method of describing a system — the following two questions are raised:

- What can be modelled using this description method?
- What can be achieved using this description?

This is illustrated on the basis of Figure 4.1. On the left-hand side we see the domains that are significant in our context, which are to be modelled in hardware description languages. Digital and analogue electronics should be unproblematic because hardware description languages were originally developed for precisely this purpose. Question marks stand next to the domains of multibody mechanics, continuum mechanics and software; the modelling of these domains using hardware description languages is investigated in this book. Furthermore, some approaches should be mentioned at this point that attempt to automatically translate further description forms into hardware description languages. The work of Maillot and Wendling [246], in which state diagrams are depicted in VHDL, is worth mentioning here. Sax *et al.* [359] transfer $MATRIX_X$ descriptions from classical control technology into VHDL-AMS. Overall, hardware description languages, and in particular VHDL-AMS, appear to be capable of serving as a general exchange format for models.

The question remains of what we can undertake using a system model in a hardware description language. This is shown on the right-hand side of Figure 4.1. Initially it is possible to specify and design using hardware description languages with the resulting models being available for documentation purposes in both cases. Furthermore, such a description can be directly simulated without any intermediate

Mechatronic Systems   Georg Pelz
© 2003 John Wiley & Sons, Ltd   ISBN: 0-470-84979-7

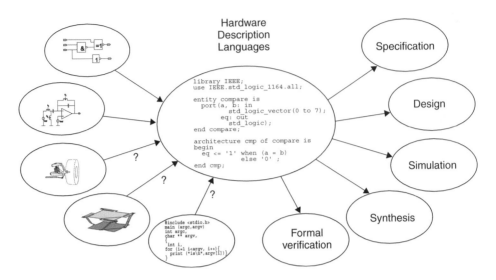

Figure 4.1   Fields of application of hardware description languages

stages, which facilitates the validation of the specifications and the verification of the designs. In the medium term formal verification or automatic synthesis of designs may also be possible, both of which currently tend to be the exclusive preserve of digital electronics.

Hardware description languages offer a whole range of advantages in relation to other approaches. For example, the problem of simulating mixed systems is moved from the simulator or programming level to the modelling level. It is thus no longer a question of implementing a tool that can execute an appropriate simulation. Instead, models have to be developed that describe the components of the system. The great advantage of this is that tried and tested simulators are available. This means that the corresponding functionalities, such as the building up and solving of equation systems, the co-simulation of digital and analogue system components or the representation of the results do not need to be re-implemented.

The second great advantage of hardware description languages lies in the fact that both the behaviour and the structure of a system or component can be formulated. Furthermore, this can occur on extremely different levels of abstraction. This allows hardware description languages to be implemented very flexibly. In particular, entire design sequences can be executed almost entirely using hardware description languages. This means that each design step primarily represents the transformation of one hardware description into another hardware description. This avoids undesirable losses due to the need to support various data formats. Furthermore, it is possible to simulate on all levels at any time and thus immediately investigate the correctness of a design step.

The most important fields of application of hardware description languages will be outlined in the sections that follow. These fields are specification, documentation, design, simulation, formal verification and synthesis. Furthermore, the syntax and semantics of hardware description languages will be represented based upon the example of the IEEE standard 1076.1 (VHDL-AMS) passed in March 1999. This lays the foundation for the subsequent chapter on modelling.

4.2 Fields of Application

4.2.1 Formulation of specification and design

A formalised circuit description on a behavioural level, such as that provided by a hardware description language, represents the precise specification and documentation of a circuit. In many cases informal paper specifications are associated with problems, for example, if certain operating states are not predicted and are thus not specified. These difficulties are avoided by using a formal, programme-like specification. With such a specification it is generally immediately clear if a system is incompletely or even contradictorily specified. Furthermore, the hardware description language is available for reference in all cases of dispute. In such a case a simulation should be capable of clearing up all doubt. Furthermore, this route automatically provides an entry into a universal design sequence. On the basis of abstract descriptions, increasingly detailed representations are developed or generated, descriptions which can be verified against one another. In this manner both the actual design problem and the problem of consistency between the textual specifications of a performance specification and the developed system can be addressed.

4.2.2 Validation of specifications and verification of designs

The use of simulations for the validation of specifications and for the verification of designs of mechatronic and micromechatronic systems is the main theme of this work. A simulator exists for virtually all hardware description languages and, for some, several simulators are even available. The simulation of digital hardware description languages has developed from logic simulation, whilst the simulation of analogue hardware description languages has developed from circuit simulation. Hardware description languages that include both digital and analogue components are represented on an appropriate 'mixed mode' simulator, which spares the user from having to think about the coupling between digital and analogue simulator cores. Nevertheless, this interface is indispensable because the simulation procedures for digital and analogue fields are very different, see Sections 2.7.2 and 2.7.3.

As an alternative to simulation we can also use the methods of formal verification in the digital field. In general, the motivation for this is that the simulation of systems almost always remains incomplete because it is not possible to play through all combinations of input values in a simulation, for reasons of running time. Formal verification makes it possible to mathematically prove the equivalence between two descriptions or the existence of certain circuit properties.

Both simulation and formal verification are normally tied to a system description in a given hardware description language. Conversely, the formulation of a system in a hardware description language often facilitates the use of appropriate tools.

4.2.3 Automatic synthesis

As indicated above, the design of a circuit often consists of an incremental refinement of a hardware description language. The theory and corresponding software tools are well developed in this field, particularly for the digital hardware descriptions. The transition from the register-transfer level to a gate net list in particular is automated as standard even now. In general, further synthesis tools are connected with this, which convert the gate net list into a standard cell layout, a gate array layout, or a programme description for a FPGA. Thus the manual part of the design sequence for digital circuits is often completed as early as the register transfer level.

4.3 Characterisation of Hardware Description Languages

At first glance a hardware description language is similar to a programming language such as C or Pascal. Models are formulated as text in a hardware description language, with a range of key words being attributed special importance. Furthermore, a predefined syntax must be adhered to. After parsing, syntactically correct models are translated into an intermediate format upon which the simulation can then be run. However, there are also important differences between hardware description languages and programming languages. For example, a programme normally runs sequentially, i.e. only one instruction is ever processed before a certain point in time. This is not acceptable for the description of hardware. All gates of a logic circuit in principle work in parallel. In hardware description languages this state is accounted for by the fact that instructions are normally processed in parallel. Certain areas of a hardware description that are reserved for sequential instructions represent the exception to this rule. In this area the typical instruments of a procedural programming language are available, such as 'if-then-else' constructs, loops or 'case' instructions.

As mentioned above, a hardware description language provides the option of describing both the behaviour and the structure of a circuit. The main difference between behaviour and structure will be explained briefly in what follows. The addition of four numbers can be unambiguously described in terms of their function as follows:

$$y = a + b + c + d; \tag{4.1}$$

The order in which the expression is evaluated is unimportant here since the commutative law for addition applies. However, if the addition is considered on the structural level then the sequence can no longer be neglected. For example, the following two alternatives exist:

$$y = (a + b) + (c + d); \tag{4.2}$$

$$y = ((a + b) + c) + d; \tag{4.3}$$

Corresponding realisations by adders are shown in Figure 4.2. It turns out that the realisation of the expression shown on the left is completed more quickly than that on the right since only two adding stages have to be run in this case.

Formulation on a behavioural level can thus significantly reduce the complexity of a circuit description. Higher operations such as addition, subtraction, multiplication, represent a few hundreds or even a few thousands of gates. Thus the readability of such a description is significantly greater than that of other circuit descriptions. Furthermore, the reuse of descriptions that were originally developed in a different context is made easier.

Finally, hardware description languages generally open up the option of considering the individual parts of a circuit in different abstractions, see Figure 4.3. Thus circuits or systems can be fully simulated if each of their modules possesses an abstract behavioural description. This initially offers an efficiency gain compared to a complete simulation of the finished design. Furthermore, as time goes on the individual blocks can be refined during the design process, until the design has achieved the required level of abstraction for the individual parts. In particular, refinements by several circuit developers can be implemented independently

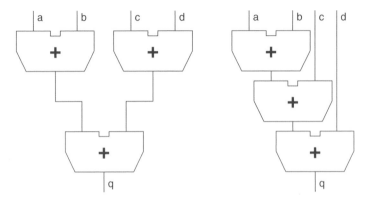

Figure 4.2 Two versions of an adder for four numbers

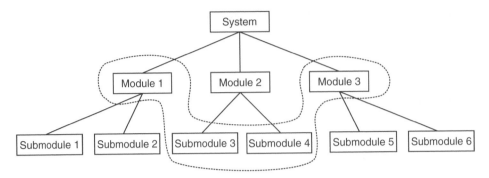

Figure 4.3   Simulation on a mixed abstraction level

of one another. Due to this 'interlacing' of the engineering work in the sense of simultaneous engineering, the design time for more complex systems can be kept within reasonable limits. Methods for the partitioning of engineering work will become increasingly important in the future because the organisational management of more complex, strongly coupled systems will increasingly be the factor that limits feasibility.

4.4 Languages

Many hardware description languages have been defined in recent years. Some of the more widespread languages were introduced by providers of design automation software. 'M-HDL' by Mentor Graphics or 'Verilog-HDL' by Cadence Design Systems are typical representatives of this group. In the analogue field the languages 'MAST' from Avant!, 'HDL-A' from Mentor Graphics, 'SpectreHDL' or 'Verilog-A' from Cadence Design Systems and 'ABCD' from Dolphin Integration S.A., are particularly worth mentioning. All these languages should be classified as proprietary hardware description languages since the associated tools could initially only be obtained from the companies in question.

A further group of hardware description languages originated from the university sector, such as 'BDS' from the University of California, Berkeley, or 'daCapo' from the University of Dortmund. However, these languages have only become widespread in the academic field. Nevertheless, because of their innovative ideas they often form the basis for commercial description languages.

A third group of languages is represented by VHDL,[2] which was initially the product of an American research programme and later became the IEEE standard 1076 as part of an expensive standardisation. The American Department of Defense, by far the biggest user in the North American area, helped the standard to make a breakthrough by making adherence to this standard a prerequisite for the

---

[2] <u>V</u>HSIC <u>H</u>ardware <u>D</u>escription <u>L</u>anguage. VHSIC = very high speed integrated circuits, American promotional program for the development of particularly powerful integrated circuits.

placement of orders. Thus all CAE providers were forced to support VHDL. Other languages were also standardised such as, for example, Verilog-HDL, which was initially designed as a proprietary language. The great advantage of such standards is that they promote the exchange of circuit descriptions and furthermore make it possible for the providers of CAE tools to exchange simulators, for example, without the reformulation of the models into another language and the significant costs associated with this. Since 1987 VHDL has been a standard for the development of digital circuits and systems, which is being continuously improved and expanded. A significant aspect of this is the expansion around analogue and mixed analogue-digital constructs. In 1999 the IEEE standard 1076.1 (VHDL-AMS[3]) was passed, which covers the full language scope of VHDL and additional constructs for the modelling of analogue processes. For an introduction to VHDL the reader is referred to the books of Ashenden [15], Pellerin and Taylor [319] and Perry [334]. With regard to VHDL-AMS, as yet there is only the provisional version of the IEEE standard 1076.1 [160] and an associated tutorial [16].

As early as 1993 VHDL and Verilog-HDL enjoyed a clear predominance in the digital field compared to other languages, see Carrol [61]. Today hardly any other languages are used in the digital field. A similar concentration will presumably also take place in the field of analogue hardware description languages.

4.5 Modelling Paradigms

4.5.1 Introduction

In the following, the most important techniques of digital and analogue modelling in hardware description languages will be described. For example, the language VHDL-AMS, which covers the most important constructs of other hardware description languages, will be considered in this connection. The aim of the descriptions that follow is to convey an impression of the modelling possibilities available using hardware description languages. However, they are not a substitute for the corresponding literature. In the following, the key words in hardware description languages are written in upper case letters and all identifiers in lower case letters. In principle this makes no difference, since in VHDL and VHDL-AMS, no differentiation is made between upper and lower case.

A VHDL model is organised into various descriptions. Every module has precisely one interface description, which in principle specifies the corresponding interface signals and their type and direction. Such a description is also called an ENTITY. For each ENTITY there is one or more ARCHITECTURE descriptions that contains the different variations of the modelling of the module. For example, in the following section three architectures will be listed for a module. For frequently used constructs it is possible to define packages, which are themselves split into

---

[3] VHDL analogue and mixed signal extensions.

an interface section (PACKAGE) and an implementation section (PACKAGE BODY). A fifth group of descriptions specifies which architectures should form the basis for a simulation. These are also called configurations (CONFIGURATION).

4.5.2 Structural and behaviour-oriented modelling

Structural modelling formulates the submodules from which a module is composed. In contrast to this, behaviour-oriented modelling describes the function and timing of the module. Let us clarify this using the example of a full adder. Hardware description 4.1 shows the interface description of a fictitious full adder in VHDL. Comments for the rest of the lines are preceded by a double minus sign. Using the LIBRARY and USE instructions a PACKAGE is first referenced, which includes the necessary types for the digital signals, e.g. std_logic. The ENTITY description mainly consists of a PORT instruction, which declares the inputs and outputs of the full adder.

```
LIBRARY IEEE;
-- IEEE Package for logic types
USE IEEE.std_logic_1164.all;

ENTITY full_adder IS
  -- two sum inputs, one Carry-In
  -- one sum output, one Carry-Out ...
  PORT (i1, i2, ci:  IN std_logic;
        sum, co:  OUT std_logic);
END full_adder;
```

Hardware description 4.1 Interface description of a full adder

The first possibility is represented by structural modelling, in which the full adder is made up of a half adder and an Or gate, see Hardware description 4.2. The timing is taken from the timing of the underlying modules.

```
ARCHITECTURE structure OF full_adder IS
...
BEGIN
...
  inst1 : half_adder(i1 ,i2 ,tc1 ,ts1);    -- Instantiation HA
  inst2 : half_adder(cin ,ts1 ,tc2 ,sum);  -- Instantiation HA
  inst3 : or_gate (tc1 ,tc2 ,co);          -- Instantiation OR
END structure;
```

Hardware description 4.2 Structural description of a full adder

The simplest form of behavioural modelling is the data flow description, in which the underlying Boolean function is merely assembled from basic functions and the calculation of the results performed after a delay. This is shown in Hardware description 4.3.

```
ARCHITECTURE data_flow OF full_adder IS
BEGIN -- Signal assignment according to Boolean function...
  sum <= i1 xor i2 xor ci AFTER 3 ns;
  co  <= (i1 and i2) or (i1 and ci) or (i2 and ci) AFTER 2 ns;
END data_flow;
```

Hardware description 4.3   Data flow description of a full adder

However, not all functions that are possible are predefined. It can also be tiresome to fully prepare the Boolean functions. In such cases it is also possible to provide a purely behavioural description, which relates input and output assignment to each other in tabular form, see Hardware description 4.4. This is based upon a so-called process, the body of which includes sequential instructions.

```
ARCHITECTURE behaviour OF full_adder IS
BEGIN
  PROCESS                                    -- Process head ...
    VARIABLE tmp : std_logic_vector(2 DOWNTO 0);
  BEGIN          -- Process body with sequential instructions ...
  WAIT ON i1, i2, ci;                 -- Wait for signal change
  tmp(2) := i1; tmp(1) := i2; tmp(0) := ci;     --Store in vect.
  CASE tmp IS                         -- Case differentiation ...
  WHEN "000" =>
    sum <= '0' AFTER 3 ns;               -- Signal assignment sum
    co <= '0' AFTER 2 ns;         -- Signal assignment Carry-Out
  WHEN "001" =>
    sum <= '1' AFTER 3 ns;               -- Signal assignment sum
    co <= '0' AFTER 2 ns;         -- Signal assignment Carry-Out
  WHEN ...
  END CASE;
  END PROCESS;
END behaviour;
```

Hardware description 4.4   Behaviour-oriented description of a full adder

4.5.3   Digital modelling

The process (PROCESS) will be explained in more detail in the following. It forms the work-horse of digital modelling. Virtually all digital relationships are modelled either directly as a process or in a form that is easy to convert into a process. The process is attributed to the parallel instructions. Thus it is processed in parallel to the other processes and the remaining parallel instructions. The body of a process contains sequential commands that are thus processed one after the other. When the processing reaches the end of the body, it jumps back to the start and thus executes an endless loop. To prevent this from causing the simulation to hang, each body must contain at least one synchronisation point in the form of an explicit or implicit WAIT instruction. Its task is to delay progress in the body of the process by an amount that depends upon its parameter. This may be based, for example, upon a fixed time period or the occurrence of a certain event. The process is executed accordingly by the performance of a sequence of instructions between two synchronisation points. Sequential instructions in VHDL are comparable to the instructions of procedural programming languages. In the following, a few processes will be described as examples.

Example: multiplexer

The first example is a multiplexer that is formulated in Hardware description 4.5 as ENTITY and ARCHITECTURE. Synchronisation takes place by means of the WAIT instruction, which interrupts the execution of the process until at least one of the signals a, b or sel has changed. Thus the body of the architecture proceeds as soon as there is a change at the inputs of the multiplexer. Then and only then can a change at the outputs be expected.

```
LIBRARY IEEE;
USE IEEE.std_logic_1164.all;      -- IEEE package for logic types
ENTITY mux IS        -- Interface description of multiplexer ...
  PORT(a, b, sel: IN std_logic;
       q          : OUT std_logic);
END mux;

ARCHITECTURE behaviour OF mux IS -- Architecture description...
BEGIN
  PROCESS                                        -- Process ...
  BEGIN
    WAIT ON sel, a, b;                  -- Wait for signal changes
    if sel = '1' then                    -- Case differentiation
      q <= a;                              -- Signal allocation
    else
      q <= b;                              -- Signal allocation
    END IF;
  END PROCESS;
END behaviour;
```

Hardware description 4.5 Behaviour-oriented modelling of a multiplexer

Example: multiplier

The next example is used to explain in more detail the various abstractions of modelling in a design sequence. The example relates to a multiplier. In its simplest form this can be described by a times sign, see Hardware description 4.6. This form of description is extremely compact, although a realisation of the circuit can consist of thousands of gates. In a second description, multiplication can be traced back to shifting and adding, as we learned multiplication at school, see

Hardware description 4.7. This corresponds with the first step in the direction of implementation. Most synthesis tools are able to translate this description into a gate circuit, which could be followed up by representation on a FPGA.

```
LIBRARY IEEE;
USE IEEE.std_logic_1164.all;     -- IEEE package for logic types
USE IEEE.std_logic_arith.all;    -- IEEE package for associated
                                    arith.

ENTITY multiplier IS
 PORT(clk: in std logic;
    a, b : IN std_logic_vector(3 DOWNTO 0);
    q    : OUT std_logic_vector(7 DOWNTO 0));
END multiplier;

ARCHITECTURE behaviour1 OF multiplier IS
BEGIN PROCESS
 BEGIN
  WAIT UNTIL rising_edge(clk);           -- Wait for rising edge
  q <= a*b;                    -- Multiplier and assign result to
 END PROCESS;
END behaviour1;
```

Hardware description 4.6 Behavioural description of a multiplier on the basis of a multiplication operation

At this point we should highlight a further point. The WAIT instructions delay the sequence up to the next active clock-pulse edge. For the architecture behaviour1 this means that the multiplication must be completed within one clock cycle. The realisation behaviour2, however, unrolls the loop over time and not spatially. Thus the calculation of the product requires at least as many clock cycles as the number of bits of the operands. In the VHDL formulation this is achieved by the fact that the loop contains a WAIT instruction.

```
ARCHITECTURE behaviour2 OF multiplier IS
BEGIN
 PROCESS
  VARIABLE pp, res : std_logic_vector(7 DOWNTO 0);
  BEGIN
   WAIT UNTIL rising_edge(clk);           -- Wait for rising edge
   res := "00000000";                 -- Initialise variable res
   FOR index IN 0 TO 3 LOOP               -- Loop index := 0 .. 3
     WAIT UNTIL rising_edge(clk);          -- Wait for rising edge
     pp := "00000000";                -- Initialise variable pp
     IF b(index) = '1' THEN          -- If bit index of b set ...
       pp((index + 3) DOWNTO index) := a;         -- Adder moved
     END IF;
     res := res + pp;                      -- Accumulate result
```

```
        END LOOP;
    WAIT UNTIL rising_edge(clk);              -- Wait for rising edge
      q <= res;                         -- Signal assignment for output
  END PROCESS;
END behaviour2;
```

Hardware description 4.7 Behavioural description of a multiplier on the basis of moving and adding

Digital signal assignment

Up until now we have based our description of a signal assignment upon an intuitive understanding, which in some cases can be deceptive. This can be clarified by looking at a simple inverter gate. The function of the inverter is quickly described. However, in some cases this does not achieve the desired result. The inverter may have a delay time of 100 picoseconds. If a pulse of one picosecond occurs at its input then we would assume in the first approximation that this pulse would be observed in the opposite polarity at the output 100 picoseconds later. However, this is not physically correct because the pulse is much too short to effect a change at the output. Before this has moved to a significant degree, the cause has disappeared again. In order to bring about this 'inert' behaviour it is necessary for each signal assignment to evaluate the right-hand side correctly and to draw up a list of current and future events. If necessary, the future events may have to be deleted again before they are realised. This is also the case, for example, if the right-hand side always produces an assignment with the same value, so that a formal assignment yields no new information for the signal. In this case we can postpone the assignment, so that no events without information content are produced. This task and others are undertaken by the so-called signal driver.

4.5.4 Analogue modelling

Introduction

We can differentiate between three classical applications of analogue modelling, see Vachoux and Berge [406]. Firstly, and self-evidently, it is implemented when the system under investigation consists wholly or partially of analogue components. But even when looking at digital systems, the consideration of an analogue environment of the circuit may still be necessary. Finally, analogue effects, such as signal delays or coupling capacitances, often cannot be disregarded especially for digital high-speed circuits.

Again, the extremely different levels of abstraction can be represented. Thus, on the purely behavioural level we can provide models based upon transfer functions or differential equations. At a lower level of abstraction, so-called macromodels are often used, which may represent the standard blocks of analogue circuit design, e.g.

operational amplifiers, comparators, etc. Such macro-models describe behaviour at the terminals, for example, in the form of a characteristic. Finally, we can also model components such as transistors, diodes, etc. using analogue hardware description languages.

Furthermore, the methodology of analogue modelling is in line with the following strategies:

Structural definition Analogue hardware description languages permit the formulation of a component as an interconnection of its subcomponents.

Behavioural definition The description of the terminal behaviour of components on the basis of mathematical equations is one of the main properties of analogue hardware description languages.

Conservative modelling Analogue hardware description languages permit the formulation of models on the basis of potential (across) and flow (through) variables, e.g. voltage and current or velocity and force, meaning that Kirchhoff's laws apply. The product of potential and flow variables is normally represented by energy. So this formulation is set up to describe energy flows.

Non-conservative modelling Non-conservative quantities can also be described, allowing block or signal flow diagrams to be formulated using hardware description languages. Often the description of an information or control flow predominates.

Table model Table models are normally based upon a piece-wise linear description, which may be smoothed for numerical reasons. These models can also be unproblematically formulated into an analogue hardware description language.

Arbitrary mixed forms Analogue hardware description languages permit the use of arbitrary mixed forms of these modelling strategies.

Using the above-mentioned modelling strategies, analogue hardware description languages thus permit the formulation of structural, physical and experimental models, so that the fundamental approaches to modelling from Chapter 2 are fully represented. The use of mathematical equations in the description of the models allows the addition of various fields to the discussion. The fields listed in Table 4.1 are particularly pertinent here, see Antao [12].

Table 4.1 Model formulation in analogue hardware description languages

Description	Field	Representation
Discrete	Time	Differential equations and algebraic equations
Continuous	Time	Differential equations and algebraic equations
Discrete	Frequency	Z-transformation
Continuous	Frequency	Laplace transformation

Now, if analogous behaviour is to be formulated in a hardware description language this normally occurs in the form of mathematical equations. In VHDL-AMS these equations are also termed simultaneous instructions. Both sides of the equation must have real values. The equations are symmetrical in the sense that swapping the left and right-hand side leads to the same results. The analogue solver is responsible for the fact that these equations are approximately fulfilled. In addition to the equations there are also the simultaneous versions of the IF, CASE and PROCEDURAL instructions, which facilitates sequential notation. Let us now clarify this using the example of a diode model.

Hardware description 4.8 shows a simple diode model in VHDL-AMS, see [160]. The division into interface and implementation, i.e. into ENTITY and ARCHI-TECTURE, also applies for the analogue model. In addition to the anode and cathode electrical connections the interface now includes a GENERIC instruction that permits the named parameter to set when the model is instanced. Furthermore, standard values are specified that are used if no further specifications are encountered during instancing. Then some electrical quantities are initially declared in the architecture such as, for example, the diode current id and the voltage ud across the diode. The threshold voltage ut is finally declared as a constant. The actual equations define the diode current id, the charge of the diode q and an additional current ic, which is found from the derivative of charge with respect to time q'DOT. Furthermore, the fact is worthy of special mention that individual equations can be allocated to a predefined accuracy group by means of the TOLERANCE instruction, so that different accuracies can be set for various equations. However, this means that no decision is anticipated regarding which criteria the simulator is to use for the evaluation of accuracy and how this is to be calculated.

```
ENTITY diode IS
  -- Parameter declaration with default values ...
  GENERIC (is0:real := 1.0e-14; tau, rd : real := 0.0);
  -- Inputs/outputs
  PORT    (TERMINAL anode, cathode: electrical);
END ENTITY diode;

ARCHITECTURE simultaneous OF diode IS
  -- Declaration of variables and constants ...
  QUANTITY ud ACROSS id, ic THROUGH anode TO cathode;
  QUANTITY q: real;
  CONSTANT ut: voltage := 0.0258;
BEGIN -- Defining equations ...
  id == is0*(exp((ud-rd*id)/ut)-1.0);
  q  == tau*id TOLERANCE "Charge";
  ic == q'DOT;
END ARCHITECTURE simultaneous;
```

Hardware description 4.8 Simultaneous behavioural description of a diode

Alternatively, a sequential description can also be provided, see Hardware description 4.9. Here the causality is specified by the assignments. However, some possibilities for sequential modelling exist such as the use of IF-THEN-ELSE constructs, CASE instructions or loops, meaning that this form of modelling also has its attraction. However formulated, the two descriptions should, however, supply the same outputs.

```
ARCHITECTURE procedural OF diode IS
 QUANTITY ud ACROSS id, ic THROUGH anode TO cathode;
 QUANTITY q: real;
 CONSTANT ut: voltage := 0.0258;
BEGIN
 p1: PROCEDURAL BEGIN -- defining assignments
    id := is0*(exp((ud-rd*id)/ut)-1.0);
    q   := tau*id TOLERANCE "charge";
    ic := q'DOT;
 END PROCEDURAL;
END ARCHITECTURE procedural;
```

Hardware description 4.9   Sequential behavioural description of a diode

Physical domains and associated quantities

When describing analogue relationships in VHDL-AMS the physical domains that can be described are not specified in advance. Rather, it is even possible to declare domains with their associated quantities subsequently. Here a differentiation is made between potentials and flows, which are declared by the keywords ACROSS and THROUGH. For electronics these may be voltage and current. Hardware description 4.10 shows the corresponding declaration as PACKAGE.

```
PACKAGE electrical_system IS
 SUBTYPE voltage IS real TOLERANCE "low_voltage";
 SUBTYPE current IS real TOLERANCE "low_current";
 NATURE electrical IS
   voltage ACROSS;    -- Potential
   current THROUGH;   -- Flow
 ALIAS ground IS electrical'reference;
 ...
END PACKAGE electrical_system;
```

Hardware description 4.10   Declaration of electrical potentials and flows

In the same manner, potentials and flows can be declared to arbitrary other domains. For translational mechanics these might be velocity and force; for rotational mechanics, rotational velocity and torque.

In a model the quantities used can be declared as either a THROUGH or an ACROSS QUANTITY. This is a real number that describes a continuous variable. Kirchhoff's voltage law is applied for potential quantities, which means that all ACROSS quantities in a closed loop add up to zero. For the flow quantities, Kirchhoff's current law applies. Thus all THROUGH quantities at a node add up to zero. In addition to the declared quantities others are implicitly defined such as, for example, q'DOT, q'INTEG and q'DELAYED(t). These denote the derivative of the quantity q with respect to time, the integral of the quantity q with respect to time and a quantity q delayed by time t. In addition to the potentials and flows it is sometimes worthwhile considering quantities that are not subject to Kirchhoff's laws. For example, in control technology signal flow diagrams or block diagrams are often considered, in which the individual quantities do not occur in pairs and furthermore have a direction. Kirchhoff's laws in particular do not apply to these quantities. In VHDL-AMS such quantities can also be used, as is demonstrated by the following example of a combined adder/integrator, see Hardware description 4.11 and [16].

```
ENTITY adder_integrator IS
 GENERIC (k1,k2: real);
 PORT     (QUANTITY in1, in2: IN real;
            QUANTITY outp: OUT real);
END ENTITY adder_integrator;

ARCHITECTURE signal_flow OF adder_integrator IS
 QUANTITY qint: real;
BEGIN -- defining equations ...
 qint == k1*in1 + k2*in2;
 outp == qint'INTEG; -- Integration
END ARCHITECTURE signal_flow;
```

Hardware description 4.11   Signal flow modelling of a combined adder/integrator

Discontinuities

In the case of mechanical models in particular, non-continuous relationships also often have to be modelled. These are illustrated in what follows based upon the example of a bouncing ball, see Hardware description 4.12 and Bakalar *et al.* [16]. Two discontinuities are considered in this model. The first of these is the start of the simulation at which the initial state is set at the first BREAK command. The second discontinuity consists of the fact that the bouncing ball reverses its velocity when the it hits a surface, i.e. at s ≤ 0. This corresponds with an elastic impact. Furthermore, the IF instruction ensures that the braking effect of air resistance acts with gravity when rising and against gravity when falling.

```
LIBRARY disciplines;              -- Reference to a package with
USE disciplines.mechanical.all;   -- the mechanical declarations
ENTITY ball IS                    -- Autonomous model,
END ENTITY ball;                  -- no connections
ARCHITECTURE simple OF ball IS
```

```
QUANTITY v : velocity;          -- Velocity
QUANTITY s : displacement;      -- Relative position
CONSTANT g : real := 9.81;      -- Gravity
CONSTANT lw: real := 0.1;       -- Air resistance
BEGIN
-- Initial conditions ...
BREAK v => 0.0, s => 10.0;
-- Detect discontinuity and invert velocity...
BREAK v => -v WHEN NOT s'ABOVE(0.0);
s'DOT == v;                     -- v = ds/dt
IF v > 0.0 USE
     v'DOT == -g - v**2*lw; -- Accel. = -Gravity - Air resist.

ELSE
     v'DOT == -g + v**2*lw; -- Accel. = -Gravity + Air resist.
END USE;
END ARCHITECTURE simple;
```

Hardware description 4.12   Modelling of discontinuities using the example of a bouncing ball

Modelling in the frequency range

In addition to modelling in the time range we can also provide a description in the frequency range. This is based upon a small-signal model, which arises as a result of the linearisation of the equations around the working point. In this model it is possible to define quantities based upon their spectra. Furthermore, predefined functions are available that effect either a Laplace or a Z-transformation. In this manner filters, for example, can be described in a very simple way.

4.6 Simulation of Models in Hardware Description Languages

In what follows the focus will again lie on the consideration of VHDL-AMS, which provides a good example of a hardware description language with digital and analogue components. Thus, we are automatically considering a mixed digital-analogue simulation. The first step is the performance of the so-called elaboration, which includes the evaluation of structural sections of the model and thus builds up a complete system model from the module instantiations. The digital section consists of a number of processes and the digital simulator core. The analogue section consists of a number of equations and the analogue solver. A necessary prerequisite for analogue solvability is that the number of equations and the number of (analogue) unknowns in the model are equal. For VHDL-AMS this is the number of THROUGH quantities, free quantities and interface quantities with the direction OUT. The actual simulation then runs in two phases. In the first phase the operating point of the system is determined. There then follows a simulation in the time, small-signal or noise range. If a model contains no quantities, then the simulation is reduced to a pure logic simulation, which corresponds with the predetermined simulation cycle in the VHDL 1076 standard, see [158] and [159]. If, on the other hand, a model does not include a digital signal, then the simulation is exclusively analogue.

The simulation cycle of VHDL-AMS should be described based upon Algorithm 4.1 below, which is formulated in pseudocode. The representation is somewhat simplified, for a complete version refer to [160].

```
Loop {
    Call to the analogue solver;
    Set current time Tc to Tn;
    If maximum time reached or no active
        processes present then simulation end;
    Bring digital signal to newest state;
    Execute active, not delayed processes
        up to the next synchronisation point (= WAIT);
    Calculate next time point of digital activity Tn;
    If Tn = Tc              -- delta time interval
        then proceed to the start of the loop;
    Execute active, delayed processes
        up to the next synchronisation point;
    Calculate next time point of digital activity Tn;
}
```

Algorithm 4.1 Simplified simulation cycle of VHDL-AMS

The simulation cycle of VHDL-AMS includes the combined simulation of analogue and digital processes and thus requires a corresponding linking of the digital and analogue solution strategies. Initially the analogue solver is called up, which in general calculates a solution up to time point T_n. However, it may be necessary for T_n to be set back to T_n' ($T_n' < T_n$), if the analogue world has produced a digital event at time point T_n'. The current time T_c is then set to T_n or possibly to T_n'. If the maximum representable time has now been reached by the time variables, or there are no longer any active processes, the simulation is ended. Otherwise the digital signals are set to the latest state and the active processes before the next synchronisation point executed. However, the execution of some of these processes is delayed. Then the next time of digital activity T_n is calculated. If T_n is equal to T_c then it is a time increment that elapses in zero time, i.e. a delta time increment. In this case execution is restarted at the start of the loop. Otherwise the delayed processes are executed and a new T_n calculated. This completes the circle and execution is recommenced at the start of the loop.

4.7 Summary

This chapter has described the opportunities of modelling in hardware description languages. It thus provides the basis for the investigation of the inclusion of software and mechanics using hardware description languages covered in the next chapter.

5

Software in Hardware Description Languages

5.1 Introduction

A whole range of methods can be listed for the joint simulation of hardware and software, which are concisely summarised by Rowson in [355]. The most important criteria here are: precision with regard to timing; simulation speed; the availability of models; and the possibility of debugging the simulated software. The simulation speed and timing precision are normally in competition with one another. The approaches described in what follows provide various compromises in this context, see Table 5.1.

The most precise, but consequently also the most expensive, simulation option is to describe the processor core in question with such accuracy that the signal timing is reproduced exactly at the connections. The software is available as information in the storage model and is processed during the simulation of hardware. This particularly exact modelling is associated with the longest running times.

We can abstract from this model and demand only that the signals at the terminals are correct at every active edge of the clock signal. This can firstly simplify the model, because for the most part the signal delays can be disregarded in a synchronously executed processor core. Furthermore, the number of simulation events is significantly reduced in comparison to the precise timing. Both lead to a significant acceleration of the simulation.

In the next step we can move to the modelling of the command set and its execution. In this procedure the values are correctly illustrated in the registers and in the memory but details such as the pipelining of instructions may be neglected. As a result a large part of the timing information is lost.

The approaches described up to this point each require suitable processor models. However, techniques exist that do not necessitate the modelling of the hardware. This is the case firstly if the communication between software and hardware runs asynchronously and the time between the communications thus plays no role. In this case it is sufficient to compile the software for the simulation workstation and

Mechatronic Systems   Georg Pelz
© 2003 John Wiley & Sons, Ltd   ISBN: 0-470-84979-7

Table 5.1   Methods of hardware/software co-simulation according to Rowson [355]

Approach	Speed in inst. / sec.	Model necessary
Exact pin timing	1–100	Yes
Cyclically precise pin timing	50–1000	Yes
Instruction level	2000–20 000	Yes
Timing disregarded	Typically limited by the hardware simulation	No

to connect to the hardware by means of a type of 'handshake'. Thus the software will be executed at the full speed of the simulation workstation. A further situation in which the timing may be neglected to a certain degree is the situation in which the execution of the software is defined in a fixed time period. Accordingly, events and new inputs are only exchanged at fixed time points. Now, if we can ensure that the software is always fast enough to conclude the calculations before the end of the current grid interval, then the timing can be disregarded. Makki *et al.* [254] suggest this for a realisation with hardware description languages, but details are not provided. Another approach is followed by van Zanten *et al.* [407] and Adamski *et al.* [3]. In this the controller core and the mechanics model — both formulated in the programming language C — are linked together and simulated jointly in the initial system investigations. The controller software is thus considered without taking into account the underlying hardware. However, this simple model of the co-simulation of hardware and software is often not adequate. The reasons for this are numerous. For example, one reason is the possible influence of an underlying real-time operating system. Also, the occurrence of further interrupts — perhaps for communication with other controllers — often frustrates the use of this variant. Finally, in some cases the aim is for the simulation to reach the speed limit, for example, in order to construct fast controllers with short calculation intervals.

Further increases in speed can only be achieved by omitting parts of the model or by the use of emulation. The latter two options will not be considered further in the following.

It is often necessary for the development of the electronics for mechatronic and micromechatronic systems to record the timing between software, electronics and mechanics with a large degree of precision, in order to thereby correctly evaluate the dynamics between the domains. A good compromise here is a simulation that reflects the temporal behaviour of the running software with regard to processor cycles. The consideration of approaches for the cyclically correct co-simulation of software, electronics and mechanics forms the focal point of this chapter. In addition to the abstractions already mentioned we must also give some thought to the realisation of the co-simulation. One possibility is to use a simulator backplane, see Gasteier and Glesner [112] or Ghosh *et al.* [118]. By contrast, the methods represented in Sections 5.2 and 5.3 increasingly point in the direction of a model transformation on the basis of hardware description languages. Finally, the method described in Section 5.4 aims at the cyclically correct coupling of software

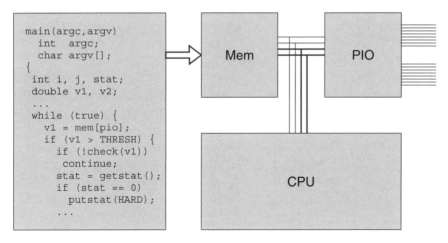

```
main(argc,argv)
    int   argc;
    char argv[];
{
    int i, j, stat;
    double v1, v2;

    ...
    while (true) {
        v1 = mem[pio];
        if (v1 > THRESH) {
            if (!check(v1))
                continue;
            stat = getstat();
            if (stat == 0)
                putstat(HARD);
        ...
```

Figure 5.1 Execution of software by the simulation of hardware

processing and hardware, but, in contrast to the backplane, this is achieved at the modelling level by hardware description languages.

5.2 Simulation of Hardware for the Running of Software

The simplest and at the same time the least efficient method for the cyclically correct co-simulation of digital hardware and software is the mere description of the hardware using hardware description languages, see for example Buchenrieder and Rozenblit [51] or Le Marrec *et al.* [218], as well as Figure 5.1. In a first approximation this takes place on the level of the blocks involved such as CPU, main memory, etc. At the start of the simulation, the modelled main memory is filled with the appropriate content so that a simulation of the hardware draws the execution of the software along with it. One such model was implemented and simulated for Motorola 68HC05 architecture. It includes behavioural models for the CPU, the main memory and a parallel interface. These models include the necessary interfaces to communicate with each other via the address and data bus. The performance of such a model lies at around 500 assembler instructions per CPU second on a SUN-Sparc 20. This is clearly too slow for the time spans in the range of seconds to be considered in mechatronics. Therefore, this approach will not be described in more detail at this point.

5.3 Co-simulation by Software Interpretation

A first step towards accelerating the cyclically correct co-simulation of hardware and software is motivated by the observation that the precise consideration of

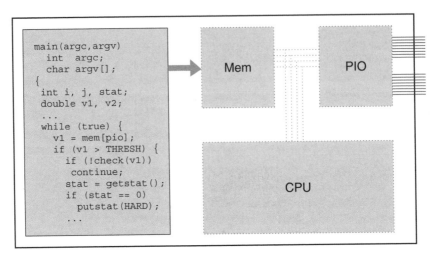

```
main(argc,argv)
  int   argc;
  char argv[];
{
  int i, j, stat;
  double v1, v2;
  ...
  while (true) {
    v1 = mem[pio];
    if (v1 > THRESH) {
      if (!check(v1))
        continue;
      stat = getstat();
      if (stat == 0)
        putstat(HARD);
      ...
```

Mem   PIO   CPU

Figure 5.2   Execution of software by simulation at controller level

bus traffic between CPU and main memory, like many other details, does not contribute significantly to the investigation of the system as a whole. Rather, it is virtually always sufficient to imitate the interface behaviour of the controller, see Figure 5.2. This facilitates a whole range of simplifications in the model. Thus it may be possible to represent the memory primarily by an array of integer numbers or bit vectors. Memory access can be formulated as access to the array. The data and address bus and the associated logic are thus dispensed with completely.

In a more precise consideration, the objective of the model in question also alters. Where before it was primarily a question of describing the hardware correctly, now such a model becomes an interpreter for the running software. This is beneficial in two respects. Firstly, the model is significantly simplified, secondly there is a considerable acceleration of the simulation. Interpretative models with various characteristics exist. For example, Gupta *et al.* [130] link an interpretative software simulator to the simulator responsible for the hardware for each simulator coupling, taking into account cyclically correct timing. Furthermore, Ecker outlines the formulation of a software interpreter in VHDL, see [92], in which precise timing is largely disregarded. Finally, Pelz *et al.* [326], [327] suggest a cyclically correct implementation of a software interpreter for the Motorola 68HC05 architecture in VHDL, which is coupled to mechanics models in hardware description languages. This approach will be described in more detail in what follows. It offers a simulation speed of around 5000 assembler instructions per CPU second on a SUN-Sparc 20. Thus the performance of the simulation lies above that of the method described in the previous paragraph by approximately an order of magnitude.

Hardware description 5.1 that follows provides an example of the description of a (fictitious) processor at interpreter level. The characteristics of the processor architecture largely relate to the register variables and the command set. The model consists of a process in which one assembler instruction is executed in each loop. At the beginning the instruction is fetched from the main memory, the Opcode is separated, and the addresses of the operands evaluated. There then follows a large CASE instruction, which serves to decode the operation in question. A few instructions are provided for each opcode, which may perform arithmetic or logical actions, set the PC in the event of jumps, calculate flags and much more.

```
ARCHITECTURE interpreter OF cpu IS
  -- Type declaration for register and main memory ...
  TYPE registers IS ARRAY (0 TO  31) OF
                    std_logic_vector(31 downto 0);
  TYPE memory    IS ARRAY (0 TO 512) OF
                    std_logic_vector(31 downto 0);
BEGIN
  cycle: PROCESS
    VARIABLE reg       : registers;                -- Registers
    VARIABLE mem       : memory;                      -- Memory
    VARIABLE pc        : natural;          -- Programme counter
    VARIABLE adr       : natural;          -- Address variable
    VARIABLE inst      : std_logic_vector      -- Instruction
                         (31 downto 0);
    VARIABLE disp      : std_logic_vector      -- Displacement
                         (31 downto 0);
    VARIABLE opcode    : std_logic_vector           -- Opcode
                         ( 7 downto 0);
    VARIABLE r3, r1, r2: natural;              -- Register adr.
    VARIABLE i8        : integer;              -- 8 bit number
    VARIABLE zflag     : std_logic;               -- Zero flag
    ...
BEGIN
    ...
    inst       := mem(pc);                -- Fetch instruction
    pc         := pc + 1;                     -- Increment PC
    opcode     := inst(31 downto 24);        -- Extract opcode
    r3         := To_Nat(inst(23 downto 16));     -- Determine
    r1         := To_Nat(inst(15 downto  8)); -- register adr.
    r2         := To_Nat(inst( 7 downto  0));-- from instruct.
    i8         := To_Int(inst( 7 downto  0)); -- Immediate Op.
    - Decode opcode ...
    CASE opcode IS
    WHEN op_add =>                                   -- Perform
    reg(r3)    := reg(r1) + reg(r2);               -- addition
    zflag      := (reg(r3) = 0) ? '1' : '0';     -- Zero flag?
    ...
  WHEN op_add_immediate =>                          -- Perform
    reg(r3)    := reg(r1) + i8;              -- addition imm.
    zflag      := (reg(r3) = 0) ? '1' : '0';     -- Zero flag?
    ...
```

```
WHEN op_sub    =>=>                                     -- Perform
   reg(r3)     := reg(r1) - reg(r2);             -- subtraction
   zflag       := (reg(r3) = 0) ? '1' : '0';       -- Zero flag?
   ...

...
WHEN op_and    =>                                       -- Perform
   reg(r3)     := reg(r1) and reg(r2);           -- logical AND
   zflag       := (reg(r3) = 0) ? '1' : '0';       -- Zero flag?
   ...

...
WHEN op_load   =>                                      -- Load reg.
   disp        := mem(pc);                     -- Determine disp.
   pc          := pc + 1;                         -- Increment PC
   adr         := To_Nat(reg(r1) + disp);   -- Determine address
   reg(r3)     := mem(adr);                              -- Load
   ...
WHEN op_store =>                                       -- Save reg.
   disp        := mem(pc);                     -- Determine disp.
   pc          := pc + 1;                         -- Increment PC
   adr         := To_Nat(reg(r1) + disp);   -- Determine address
   mem(adr)    := reg(r3);                       -- Store in mem.
   ...
WHEN op_branch_on_zero =>                            -- Jump command
   IF    (zflag = '1') THEN                         -- If flag = 1
      disp        := mem(pc);                  -- Determine disp.
      pc          := pc + 1;                      -- Increment PC
      adr         := pc + To_Nat(disp);     --    Determine address
      pc          := adr;                              -- Set PC
      ...
   END IF;
   ...
WHEN others =>
   -- Unknown opcode ...
   ASSERT false REPORT "illegal instruction"
      SEVERITY warning;
   WAIT;
END CASE;
END PROCESS;
END ARCHITECTURE;
```

Hardware description 5.1 VHDL description of a simple processor as software interpreter

5.4 Co-simulation by Software Compilation

5.4.1 Introduction

The approach described in the previous section interprets software during the running time in order to process it. This generates a considerable cost to be paid during simulation. The better alternative is to shift the compilation cost from the running time to a pre-simulation stage. This generally means that two versions of the software exist. One is compiled for the simulation workstation, the other is compiled for the processor on which it is to run in the system. Now, if the software exists in a higher programming language and we are only interested in the function and not in the timing, then the differences between the processors do not play a significant role. The prerequisite for this is that the software always calculates a certain result within a predetermined time period. A whole range of approaches to HW/SW co-simulation are based upon this principle such as, for example, the work of Becker *et al.* [21] or Thomas *et al.* [399].

We can expand upon this methodology so that cyclically correct timing is also taken into account. However, to achieve this we have to make a detour in the modelling. In a first step the assembler or machine programme is compiled into a C routine that both reflects the functionality and correctly takes into account the timing of the software execution on the basis of the clock cycles of the target processor. Živojnović and Meyr show this in [438] for pure digital electronics, with both software and electronics being described in C modules so that these only have to be linked together. Pelz *et al.* expand upon this approach in [328] based upon a compiled co-simulation of software, electronics and mechanics by implementing an appropriate synchronisation between simulator and software model in a hardware description language. Here the representation of the assembler programme in C is automated by a compiler based upon a disassembler. Overall this method can also be regarded as a modelling of software, see Figure 5.3.

In what follows this approach of representing system software in C routines and linking it into a simulator on the basis of hardware description languages will be investigated in further detail.

5.4.2 Software representation

In a first stage, the system software should be represented in a C routine that takes into account both the function and timing on the level of machine instructions. In

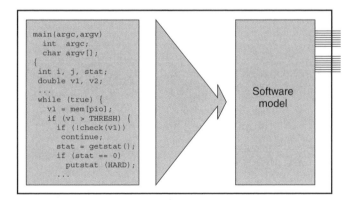

Figure 5.3 Execution of software by modelling at software level

order to subsequently bring about synchronisation, it must be possible to leave the routines at any desired points and re-enter them again later; they must therefore be 're-entrant'.

Furthermore, it is necessary that they have a memory so that the applicable system state can be held in the form of a context. Such a context thus includes the registers of the underlying processor and the complete main memory. Furthermore, a second context is saved in parallel so that the synchronisation — as described more precisely later — can refer back to an old state.

The basic idea is now to store short blocks of C instructions, which each represent an assembler instruction, one after the other in a routine. The sequence of C blocks thus corresponds with the sequence of assembler instructions, so that sequential progress through the assembler instructions corresponds with sequential progress through the C blocks.

A C block for an assembler instruction in principle contains the following components:

- Execution of the operation, e.g. for arithmetic and logical operations.

- Setting of the flags, depending upon operation.

- Setting the programme counter, normally by an increment based upon the byte number of the operation, or in the event of jumps an addition (relative) or an assignment (absolute).

- Protecting the return address on the stack in the event of subprogramme calls.

- Addition of the number of required cycles on the cycle counter.

- Calculation of the current time from the cycle counter.

- Control of the debugger.

- Details of the representation will be described in Section 5.4.4 on the basis of an example.

5.4.3 Synchronisation

Introduction

The synchronisation between hardware and software serves to effect the correct chronological sequence of events in the software model and hardware model in the simulator. A significant prerequisite for a simple and efficient solution is that the simulation of the hardware runs in a linear manner and at most is delayed only now and then. All other strategies would have an effect deep within the logic or circuit simulator used, thus shifting the problem from the modelling level to the tool level, which would often rule out solutions based upon commercial simulators.

In order to achieve this the software should run for a defined time span. This is effected by calling up an external C routine from the hardware model. With regard to the timing of the return of the software, the question is raised as to whether the sequence of load or store instructions includes reference to the I/O ports, i.e. whether it wants to exchange data from within itself with the hardware. If this is the case then the processing of the software is interrupted immediately. Otherwise the software runs until the predefined time point. Upon return, the C routine informs the hardware of the time point t that it reached. Since the software has run in zero time from the point of view of the hardware, the hardware should now be simulated up to time point t so that time equality exists between software and hardware, and thus data can be exchanged if necessary. However, the sequence described up until now only functions as long as no interrupt is triggered. In the event of an interrupt occurring, the state of the software is initially brought to the time point at which the interrupt occurred. Then synchronisation occurs and the programme counter is set to the interrupt vector, whereupon the normal sequence can once again be resumed. The forms of synchronisation described thus far will be considered in more detail in what follows.

Synchronisation without interrupt

Let us initially assume that no access to I/O ports has occurred during the processing of the software, see Figure 5.4. Before the software can once again proceed for a certain period of time, a synchronisation must take place. This means primarily that we wait until the hardware has also been simulated up to the time point at which the software currently stands. When the software and hardware show the same value for time, the software can once again proceed and the described procedure runs from the start.

Figure 5.5 illustrates the case of access to the I/O port. Here the occurrence of a corresponding load or store command leads to the software sequence being

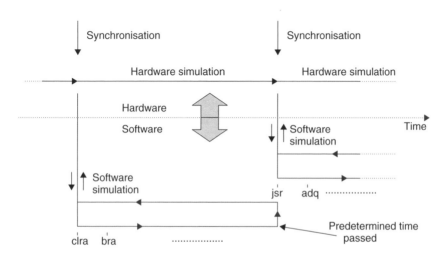

Figure 5.4 Synchronisation between hardware and software after the time allotted for the software has elapsed

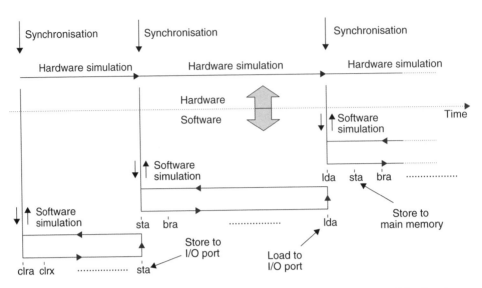

Figure 5.5   Synchronisation of software and hardware after the occurrence of load and store instructions (lda and sta) relating to the I/O ports

interrupted. Then we again wait until the hardware has reached the current time of the software. At this point the appropriate values can be exchanged between hardware and software. Then the software is restarted.

Synchronisation after an interrupt

This case occurs if the software has been executed up until time point t and it is found during the hardware simulation that an interrupt has been triggered at time point t' < t, that has invalidated the current progress of the software simulation, see Figure 5.6. The problem is solved in two stages. In the first stage the software has to be brought back to its state at the time of the interrupt t'. We first jump back to the old state that is stored at the start of every software operation. This is also called a time-warp in the literature on the general coupling of simulators, see the work of Jefferson [168] and [169]. Then the software is simulated up until the time of the interrupt. We can think of this as a type of 'replay' of a sequence that has already played out in the past. After the replay the software shows the precise state at time point t'. A synchronisation point is then inserted here, which permits the interrupt to be taken into account at exactly the right time. Then the software simulation begins again from the instruction that refers to the interrupt vector.

5.4.4 Example of software modelling

The representation of the software shall be explained on the basis of an example in what follows. Programme 5.1 shows parts of an assembler programme and

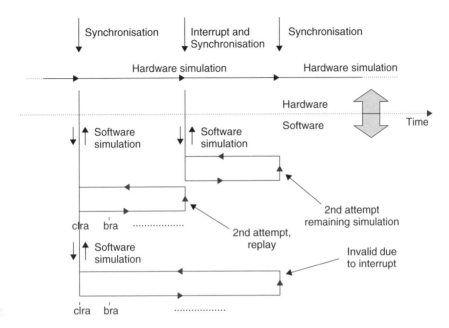

Figure 5.6 Synchronisation of software and hardware after the occurrence of an interrupt

Programme 5.2 shows the corresponding C routine which was automatically generated. Both the assembler instructions in question and the context of the C routine
are compatible with the architecture and the command set of the Motorola 68HC05
microcontroller.

```
PORTA: EQU $0010 ; Declaration of PORT A as address
PORTB: EQU $0001 ; Declaration of PORT B as address
PORTC: EQU $0002 ; Declaration of PORT C as address
PORTD: EQU $0003 ; Declaration of PORT D as address

       org $100  ; Position in the memory: 0100 Hex
START:           ; Start label
       lda PORTA ; (load A) Load port A in accumulator
       jsr SRX   ; (jump subroutine) Execute subroutine SRX
       bra SRY   ; (branch) Branch to label SRY
       ...
SRY:             ; Label SRY
       ...

       org $200  ; Position in the memory: 0200 Hex
SRX:             ; Label of the subroutine SRX
       ...
```

Programme 5.1 Excerpt from assembler programme

Upon its call up, the fundamental sequence of the C routine initially rests upon
determining whether this is the first time the routine has been run. If so, a whole

range of initialisations are necessary, such as, for example, filling the memory with the programme, resetting the register and jumping to the first instruction.

If the C routine has been called before, the correct context must first of all be created. If it is a replay the old stored context is activated by exchanging (exchange_context) with the current context. Then the old context is always protected by copying (copy_context). The jump to the hub brings about a jump to the instruction referred to in the programme counter of the current context.

The lda, jsr and bra instructions from the assembler programme can also be found in the C routine. There are called by labels (1256, 1258, 1261), which permit jumping to the instructions using the goto command.

First the lda should be considered more closely. Depending upon the targetted address this command fetches a value from the memory or from a port and stores it in the accumulator. First a routine is called up for this instruction, which controls the debugger and thus permits it to visualise the software sequence, indicate values, and control the software sequence by means of breakpoints. The user interface of the debugger is shown in Figure 5.7. The next instruction decides whether the

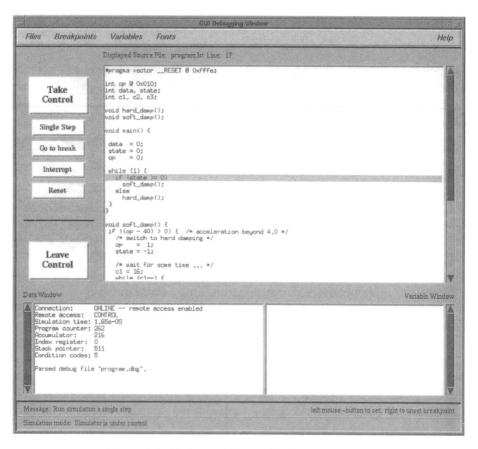

Figure 5.7   Software debugger for virtual hardware

address given in the direct addressing is a port represented in the memory area or a memory location in the main memory. In the first case the addressed port is accessed via the routine `fetch_io`. In the second case the accumulator `c1->ac` is set to the value of the memory cell at which the byte points to the opcode. It should also be mentioned that the pointer `c1` points to the current context. A type declaration of the context is located at the start of the C routine. Then the cycle counter is incremented by 3 and the programme counter by 2. Finally, the affected flags are updated and the current time `t_cur` calculated. The two other commands shown are processed in a similar manner.

The `jsr` instruction describes the call of a subroutine, so that the return address is initially stored on the stack in two bytes. Then the address of the subroutine is calculated from the two bytes following the opcode and entered into the programme counter. Then the cycle counter is incremented and the current time calculated. Finally there is a jump to the label `hub` at which the large `switch` instruction initiates a jump to the correct label. This diversion is necessary because in C it is not generally possible to jump to a variable destination by means of a `goto` command.

Finally, the `bra` instruction includes the calculation of a relative jump, which can also be in a backwards direction. The second byte of this instruction — the width of the jump — should thus be regarded as a signed number, which is expressed in the appropriate C instruction. After the normal incrementation of the cycle counter and the calculation of the time the actual jump again takes place via the hub.

```
typedef struct context {
  /* Programme counter (pc), Accumulator (ac), Index register
     (ix), Stack pointer (sp), Flag register (cc), Cycle
     counter (cyc), Main memory (m) ...*/
  unsigned int ac, ix, sp, pc, cc, cyc, ...;
  unsigned int m[MEMORYSIZE];
  ...
} CONTEXT;
static CONTEXT con1, con2, *c1=&con1, *c2=&con2;

software_sim(t_start, t_stop ... ) ... ; {
  if (t_start > 0.0) {
    if  (t_start ss< t_cur_old) {
      /* t_cur_old = Time when routine was last left,
         Replay! ... */
      exchange_context(&c1,&c2);
      ...
    }
    copy_context(c1,c2);
    ...
    goto hub;
  }
  else {
```

```c
/* Start time = 0, first call:
   Initialise debugger, logger, context etc.
   Fill the main memory with the programme */
c1->m[256] = 182; c1->m[257] = 16;
c1->m[258] = 205; c1->m[259] = 1; c1->m[260] = 20;
c1->m[268] = 32;  c1->m[269] = 3;
...
/* Initialise context ... */
c1->pc = 256*c1->m[MEMORYSIZE-2] + k1->m[MEMORYSIZE-1];
c1->ac = 0; c1->ix = 0; c1->sp = 511; c1->cc = 0;
...
goto hub;
/* Assembler programme in C ... */
10256:      /* lda, Load Accumulator, direct addr. */
  debugger(...);                    /* Control debugger */
  if (is_io(c1->m[c1->pc+1]))/* IO or main memory? */
    c1->ac=fetch_io(c1->m[c1->pc+1]);/* IO access */
  else
    c1->ac=c1->m[c1->m[c1->pc+1]];/* Main memory access */
  c1->cyc+=3; c1->pc+=2;     /* Increment cyc, pc */
  set_flags(...);                   /* Update the flags */
  t_cur=c1->cyc*CYCTIME;          /* Update the time */
10258:         /* jsr, Jump Subroutine, ext. addr. */
  debugger(...);                   /* Control debugger */
  c1->m[c1->sp--]=(c1->pc+3)%256;/* Protect return */
  c1->m[c1->sp--]=(c1->pc+3)/256;/*   address on stack */
  c1->pc=256*k1->m[c1->pc+1]+c1->m[c1->pc+2];/* Set pc */
  c1->cyc+=5;                      /* Increment cyc */
  t_cur=c1->cyc*CYCTIME;          /* Update time */
  goto hub;                       /* Initiate the jump */
10261:      /* bra, Branch, relative addressing */
  debugger(...);                  /* Control debugger */
  c1->pc=c1->pc+2+c1->m[c1->pc+1]>127 ?/* Calculate rel. */
  (-(256-c1->m[c1->pc+1])):(c1->m[c1->pc+1]);/* jump */
  c1->cyc+=3;                      /* Increment cyc */
  t_cur=c1->cyc*CYCTIME;           /* Update time */
  goto hub;                        /* Initiate the jump */
...
hub:
switch(c1->pc) {
  case 256: goto 10256;
  case 258: goto 10258;
  case 261: goto 10261;
  ...
}}}
```

Programme 5.2 Simplified software model in programming language C

The main task of synchronisation is to act as an interface between software and external hardware. Externally it adopts the connections of the processor. Internally the C routine is called up. In accordance with the preceding representation of the synchronisation algorithms, a formulation in a hardware description language will now be represented, see Hardware description 5.2. The language used here is MAST (Avant!) because the research work in question, see [328], was performed in this language.

The majority of the synchronisation lies in a WHEN instruction, the body of which is executed if its condition is true. It is thus largely comparable with a process in VHDL. In the body there is initially an interrogation of the interrupt line to determine whether a replay is necessary. This is performed if necessary, and then the actual execution of the software takes place. Upon return from the C routine the software reports that it was able to simulate until time point t_cur. Then an event at time t_cur is initiated upon the softsync signal. When this occurs, software and hardware are synchronised. Thus data can be exchanged and a new software operation started. This is taken into account accordingly by the WHEN instruction.

```
template m6805 ...                        # Interface description

  . . .
{
  state time t_cur              # Current software time upon return
  state time t_old        #  Start time of the last software call
  state time step     #  Desired length of the software operation
  state logic_4 softsync    #  Carries events for synchronisation
  foreign software_sim                       # External C routine
  . . .

  # If simulation beginning or event at the softsynch
  # variable or active edge on the interrupt line ...

  when (time_init | event_on(softsync) |
        (event_on(interrupt)&(interrupt==14_0)) ) {

    if (event_on(interrupt)&(interrupt==14_0)) {     # Interrupt!
      # Replay, time supplies the current time
      (t_cur, ...) = software_sim(t_old,time,...)     # C-Routine
      . . .
    }
    . . .
    (t_cur, ...) = software_sim(time,time+step,...)  # C-Routine
    schedule_event(t_cur, softsync, 14_1) )     # softsync event
    t_old = time                        # Save old start time
    . . .

  }
}
```

Hardware description 5.2 Simplified description of the synchronisation between hardware and software in the hardware description language MAST

5.4.5 Debugging of software

The visualisation of software cannot be achieved in a worthwhile manner using the tools of an electronics or mechanics simulator. Ideally, the tools used for pure software development would be used. Such debuggers show the instruction currently being executed and the content of the variables. Furthermore, it is possible to act upon the sequence of the software by setting breakpoints and then investigating particular points in steps. It should also be possible to change the value of the variables during execution.

However, we are dealing with software that is run on virtual hardware. Furthermore, feedback effects from electronic and possibly mechanical system components, also have to be taken into account. Such a debugger has been developed, see Pelz *et al.* [328], and correspondingly incorporated into the software model. Figure 5.7 shows the user interface that has been developed for this.

The two buttons 'Take Control' and 'Leave Control', which allow us to take over the control of the simulation or leave it again, are of primary importance. In control mode we can move forward in 'Single Step' mode or proceed directly to the next halt point 'Go to break'. An 'Interrupt' interrupts such a sequence, whilst 'Reset' restores the original state. In the top window the system programme is displayed at assembler or programming language level. Clicking on a line sets or recalls a break. The bottom left window shows the most important system information, and particularly the current content of the register, whilst the bottom right window shows the variable contents.

5.5 Summary

In this chapter the inclusion of software using hardware description languages was investigated. Using the results obtained we can now look at systems that incorporate software components in addition to electronics and other domains. Significant features are the cyclically correct management of software operation on a controller, efficiency as a result of the compiled simulation of the software, and the options of linking in a debugger for the visualisation and control of the simulation process. Using the methods for the modelling of mechanics in hardware description languages, dealt with in the next chapter, yields a universal modelling process for mechatronic and micromechatronic systems that can be executed directly upon available commercial simulators.

6

Mechanics in Hardware Description Languages

6.1 Introduction

The objective of this section is to highlight the most important strategies for obtaining the equations of motion for mechanical components and systems and to clarify the options for their subsequent representation in hardware description languages. Both direct formulations of symbolic equations and indirect formulations based upon the parametric calculation of the system matrices will be considered. The latter is often also called the solution using numerical equations.

The use of hardware description languages for the modelling of mechanics also implies that the solution of the mechanical equation takes place using the solver of the circuit simulator. Naturally, solvers are generally optimised for various domains. For electronics the focus tends to be upon the management of a large number of degrees of freedom, whereas in mechanics numerical problems with a large number of constraints require particular attention. On the other hand, the example of the classical multibody simulator ADAMS shows that this contrast is not irreconcilable, see Orlandea *et al.* [304] and [305]. The numerics of ADAMS is largely based upon procedures that are also used in circuit simulation. We should mention at this point that the equation system is not formulated using a minimum number of equations according to the degrees of freedom. Rather, each individual equation is entered into an overall system. Thus the resulting system matrix is sparse and can be processed using 'sparse matrix' techniques. The numerical integration takes place using the Gear procedure that is also commonly used in circuit simulation.

In mechanics — as in electronics — we can differentiate between various abstractions. Multibody mechanics and continuum mechanics are examples. According to Schiehlen [361] a multibody system is characterised as a collection of rigid and/or elastic bodies with inertia as well as springs, dampers and servo motors without inertia. These are connected together by rigid bearings, joints or suspensions. Friction and contact forces can also be included if necessary. This corresponds with

Mechatronic Systems   Georg Pelz
© 2003 John Wiley & Sons, Ltd   ISBN: 0-470-84979-7

modelling using concentrated parameters and is thus comparable with a circuit made up of components.

In some cases the abstraction of a continuum to the discrete elements of a multi-body system is not suitable for the solving of the envisaged problem. For example, this is the case if the exact deformation of an elastic body contributes significantly to the system behaviour. In this case models need to be created and formulated in hardware description languages that adequately describe the continuum with its distributed parameters. Both multibody mechanics and continuum mechanics will be considered in the following.

6.2 Multibody Mechanics

6.2.1 Introduction

When modelling multibody mechanics using hardware description languages, we first have to raise the question of the perspective from which the system is to be considered. One option focuses upon the system level, the other upon the component level with the system models being generated by connecting component models together. The first method is called system-oriented modelling and the second method object-oriented modelling.

The first option has the advantage that equations of motion can be created using standard engineering methods. Furthermore, we have access to a greater system knowledge during the modelling, which can be beneficial. However, one problem is that this type of consideration opposes one of the most important basic philosophies for the development of electronics. In this field it is generally sufficient to model only the fundamental components and to develop complex systems from these. Further modelling is normally not necessary during the development of electronic systems.

As an alternative to this we can use object-oriented modelling to describe the standard components — for example bodies, springs, dampers, joints, etc. — and put these submodels together into a system model. Information about this system, such as, for example, a favourable selection of generalised coordinates, is in principle not available and thus cannot be used for the simplification or acceleration of the model. However, the building of a system model can be considerably simplified if the basic models that are required are available.

In the following we will consider how it is possible to obtain equations of motion for multibody systems, see for example, Dankert and Dankert [79], Greenwood [125], Hiller [144] or Nikravesh [299] for the basic principles shown. Multibody systems typically include the following components:

- Particles with translational inertia.

- Rigid bodies with translational and rotational inertia.

- Suspensions and joints that limit the movement of individual particles and bodies in relation to one another.

- Coupling elements, e.g. springs, dampers, servo motors, etc.

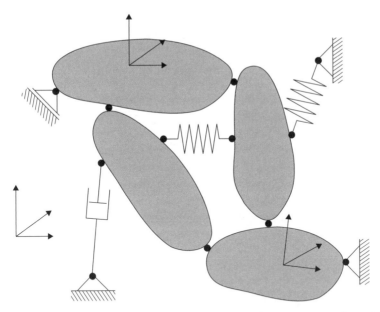

Figure 6.1 Multibody system with four bodies, springs, dampers, suspensions, joints, and inertial and body-related frames of reference

In the consideration of the structure of a multibody system, an abstracted description such as that given in Figure 6.1 is generally sufficient. Decisive factors are the topography of the system and the parameters of the individual elements, such as mass, centre of gravity, moments of inertia with respect to the main axes or the point of application of forces.

For the consideration of point-shaped masses we start from Newton's second law, which identifies the product of mass m and acceleration in the x, y, and z direction a_x, a_y, a_z of a particle with the forces F_x, F_y, F_z acting upon it:

$$F_x = ma_x, \qquad F_y = ma_y, \qquad F_z = ma_z \tag{6.1}$$

Let us now consider a system of N particles. These may be subject to additional limitations to their movement, so-called constraints. This state of affairs can be taken into account by the introduction of the so-called reaction forces, which ensure that the constraints are adhered to. The total force acting upon a body is divided into two components, the force applied from outside F_i^e and the reaction force F_i^r. In total this yields the following equation system:

$$m_i a_{ix} = F_{ix}^e + F_{ix}^r$$
$$m_i a_{iy} = F_{iy}^e + F_{iy}^r \qquad (i = 1, 2, \ldots, N) \tag{6.2}$$
$$m_i a_{iz} = F_{iz}^e + F_{iz}^r$$

This can be formulated as a vector equation as follows:

$$m_i \mathbf{a_i} = \mathbf{F_i^e} + \mathbf{F_i^r} \tag{6.3}$$

For the sake of simplicity we can also describe the cartesian coordinates of the
first body by (x_1, x_2, x_3), those of the second body by (x_4, x_5, x_6) and so on. If we
also term the masses of the kth body $m_{3k-2} = m_{3k-1} = m_{3k}$ and set $a_i = \ddot{x}_i$ for the
accelerations, then the equations of motion can be formulated by the following set
of equations:

$$m_i \ddot{x}_i = F_i^e + F_i^r \qquad (i = 1, 2, \ldots, 3N) \tag{6.4}$$

If the movement of the particle is not restricted then the reaction forces are neg-
ligible. This yields a system of 3N second-order differential equations, which is
generally nonlinear. This equation system can in general only be solved numerically, i.e. as part of a simulation.

The constraints between the particles are characterised by a set of M independent
constraint equations:

$$f_j(x_1, x_2, \ldots, x_{3N}, t) = 0 \qquad (j = 1, \ldots, M) \tag{6.5}$$

So $3N + M$ equations are available for the solution of the same amount of variables.

However, the use of cartesian coordinates is not always favourable. In many
cases cylindrical, spherical, elliptical, parabolic or other coordinates are beneficial. For this reason we will now move to the so-called, generalised coordinates
$q_1, \ldots, q_n$. These permit a formulation that is better suited to the problem. Furthermore, under certain conditions the generalised coordinates can be selected so that
the constraint equations are dispensed with completely, considerably simplifying
the drawing up and calculation of the equations of motion. This is possible if all
constraints are holonomous, i.e. they relate exclusively to the possible geometric
positions of the bodies or can at least be put into such a form. Regardless of the
selection of coordinates, the number of degrees of freedom of the system in principle remains constant. It corresponds with the number of independent coordinates
minus the number of independent constraint equations.

A small example, see Greenwood [125], should clarify the relationship between
cartesian and generalised coordinates, see Figure 6.2.

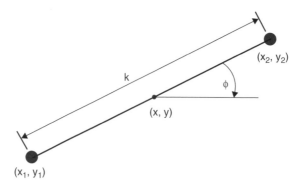

Figure 6.2   Description of the position of two particles joined by a mass-free rod

Two mass points in the plane are rigidly joined together by a mass-free rod. Their position is determined by two pairs of cartesian coordinates (x_1, y_1) and (x_2, y_2). The condition induced by the rod can be described by the following equation

$$(x_2 - x_1)^2 + (y_2 - y_1)^2 - k^2 = 0 \qquad (6.6)$$

We therefore have four cartesian coordinates and a bond equation, thus a total of three degrees of freedom. In principle, however, the configuration of the two particles can be described by the following generalised coordinates:

$$q_1 = x \quad \text{coordinate of the mid-point of the rod}$$

$$q_2 = y \quad \text{coordinate of the mid-point of the rod}$$

$$q_3 = \text{angle } \phi \text{ of the rod.}$$

The fourth coordinate q_4 would be the length of the rod, which is, however, constant. So the associated bond equation

$$q_4 = k \qquad (6.7)$$

is trivial and can be disregarded. There thus remain three coordinates without a further bond equation, i.e. three degrees of freedom. Only by this formulation in generalised coordinates can we thus omit the consideration of constraint equations in holonomous systems. In the following we will consider exclusively holonomous systems.

It is often worthwhile going over to the generalised coordinates, which — as is the case for pure cartesian coordinates — represent the configuration of the system, i.e. the position of all particles. Coordinate transformations permit the conversion between generalised and cartesian coordinates:

$$x_1 = x_1(q_1, q_2, \ldots, q_n, t)$$

$$x_2 = x_2(q_1, q_2, \ldots, q_n, t)$$

$$\vdots \qquad\qquad\qquad\qquad (6.8)$$

$$x_{3N} = x_{3N}(q_1, q_2, \ldots, q_n, t)$$

The transition to generalised coordinates requires that forces acting upon the system from outside are also present in generalised form. Whereas the forces in cartesian coordinates can be simply split up into their x, y and z components, things are more complicated in this case. For example, forces acting upon angular coordinates become moments. The conversion rule for the generalised force Q_i is naturally also based upon the coordinate transformation x_j and looks like this:

$$Q_i = \sum_{j=1}^{3N} F_j \frac{\partial x_j}{\partial q_i} \qquad (i = 1, 2, \ldots, n) \qquad (6.9)$$

If we now move on from particles to rigid bodies we now have to consider the moments of inertia in addition to the translational inertia. These are described by the underlying Euler equations for the rigid body K_j:

$$I_{j1}\dot{\omega}_{j1} - (I_{j2} - I_{j3})\omega_{j2}\omega_{j3} = M_{j1}^e + M_{j1}^r$$

$$I_{j2}\dot{\omega}_{j2} - (I_{j3} - I_{j1})\omega_{j3}\omega_{j1} = M_{j2}^e + M_{j2}^r \qquad (6.10)$$

$$I_{j3}\dot{\omega}_{j3} - (I_{j1} - I_{j2})\omega_{j1}\omega_{j2} = M_{j3}^e + M_{j3}^r$$

In matrix form these equations look like this:

$$\mathbf{I_j}\dot{\boldsymbol{\omega}}_j + \boldsymbol{\omega}_j \times (\mathbf{I_j}\boldsymbol{\omega}_j) = \mathbf{M_j^e} + \mathbf{M_j^r} \qquad (6.11)$$

where $\mathbf{M_j}$ represents the applied and reactive torque vectors, $\mathbf{I_j}$ represents the tensors of the moment of inertia and $\boldsymbol{\omega}_i$ represents the angular velocities with respect to the three principal axes of the rigid body K_j.

6.2.2 System-oriented modelling

In system-oriented modelling two classical approaches can be distinguished, the synthetic and the analytical, see for example, Kreuzer [207]. In the synthetic methods we first draw up the Newton and Euler equations for each body. The connections between bodies, e.g. joints, give rise to constraining forces, and the elimination of these converts the Newton/Euler equations into equations of motion. The analytical approach, on the other hand, is associated with the name Lagrange and starts from an energy formulation. This is rearranged directly into equations of motion without the constraining forces being considered.

Both approaches will be described in the following in a formulation using generalised coordinates. In addition to the above-mentioned approaches there is also a range of further options, which are briefly described and compared by Kane and Levinson in [177]. It should not go unmentioned that the equations that result from the various approaches are ultimately the same. However, they are obtained at a different level of complexity. The formulation is also of varying suitability for the subsequent numerical simulation.

Newton–Euler approach

The Newton–Euler approach, see also Kreuzer and Schiehlen [208], should—just like the Lagrange approach described subsequently—be represented in a formulation using the generalised coordinates $q_1, \ldots, q_n$. From these the velocities should be determined for each body K_j in the x, y and z coordinates:

$$\begin{bmatrix} v_{xj} \\ v_{yj} \\ v_{zj} \end{bmatrix} = \begin{bmatrix} \dfrac{\partial x_j}{\partial q_1} & \dfrac{\partial x_j}{\partial q_2} & \cdot & \cdot & \dfrac{\partial x_j}{\partial q_n} \\[2mm] \dfrac{\partial y_j}{\partial q_1} & \dfrac{\partial y_j}{\partial q_2} & \cdot & \cdot & \dfrac{\partial y_j}{\partial q_n} \\[2mm] \dfrac{\partial z_j}{\partial q_1} & \dfrac{\partial z_j}{\partial q_2} & \cdot & \cdot & \dfrac{\partial z_j}{\partial q_n} \end{bmatrix} \begin{bmatrix} \dot{q}_1 \\ \dot{q}_2 \\ \cdot \\ \cdot \\ \dot{q}_n \end{bmatrix} + \begin{bmatrix} \overline{v}_{xj} \\ \overline{v}_{yj} \\ \overline{v}_{zj} \end{bmatrix} \tag{6.12}$$

In more compact form the same situation can be formulated as follows

$$\mathbf{v_j} = \mathbf{J_{Tj}}\dot{\mathbf{q}} + \overline{\mathbf{v}}_j \tag{6.13}$$

where both the translational Jacobi matrix $\mathbf{J_{Tj}}$, and the local velocities $\overline{\mathbf{v}}_j$ depend only upon q and t. The cartesian accelerations $\mathbf{a_j}$ of the jth body are calculated as follows:

$$\mathbf{a_j} = \mathbf{J_{Tj}}\ddot{\mathbf{q}} + \overline{\mathbf{a}}_j \tag{6.14}$$

For the local accelerations $\overline{\mathbf{a}}_j$ it is again true that they depend only upon q and t.

In a similar manner we move from the generalised coordinates to the angular velocities $\boldsymbol{\omega}_j$ and angular accelerations $\boldsymbol{\alpha}_j$ of each rigid body K_j:

$$\boldsymbol{\omega}_j = \mathbf{J_{Rj}}\dot{\mathbf{q}} + \overline{\boldsymbol{\omega}}_j \tag{6.15}$$

$$\boldsymbol{\alpha}_j = \mathbf{J_{Rj}}\ddot{\mathbf{q}} + \overline{\boldsymbol{\alpha}}_j \tag{6.16}$$

where the rotational Jacobi matrix $\mathbf{J_{Rj}}$ and the local angular velocities $\overline{\boldsymbol{\omega}}_j$ and angular accelerations $\overline{\boldsymbol{\alpha}}_j$ again depend exclusively upon q and t.

In the next step we draw upon Newton and Euler equations for each body:

$$m_i \mathbf{a_i} = \mathbf{F}_i^e + \mathbf{F}_i^r \tag{6.17}$$

$$\mathbf{I}_j \dot{\boldsymbol{\omega}}_j + \boldsymbol{\omega}_j \times (\mathbf{I}_j \boldsymbol{\omega}_j) = \mathbf{M}_j^e + \mathbf{M}_j^r \tag{6.18}$$

Using the transformation of generalised coordinates into (angular) velocities and (angular) accelerations of the individual body, as described above, these equations can now be formulated exclusively in the form of generalised coordinates. However, these are the same for all bodies, which means that the bodies can be linked together in this manner. The resulting system of equations takes the following form:

$$\overline{\mathbf{M}\mathbf{J}}\ddot{\mathbf{q}} + \overline{\mathbf{k}} = \overline{\mathbf{p}}^e + \overline{\mathbf{p}}^r \tag{6.19}$$

Where the 6k × 6k matrix $\overline{\mathbf{M}}$ takes the form

$$\overline{\mathbf{M}} = \mathrm{diag}(m_1\mathbf{E}, \ldots, m_k\mathbf{E}, \mathbf{I_1}, \ldots, \mathbf{I_k}) \tag{6.20}$$

and forms a block diagonal matrix of masses and inertia tensors. The k denotes the number of bodies. The 6k × n matrix $\overline{\mathbf{J}}$ is the global Jacobi matrix and consists of a stack of k translational and k rotational 3 × n Jacobi matrices of the individual bodies, where n is the number of generalised coordinates:

$$\overline{\mathbf{J}} = [\mathbf{J_{T1}^T}|\cdots|\mathbf{J_{Tk}^T}|\mathbf{J_{R1}^T}|\cdots|\mathbf{J_{Rk}^T}]^T \tag{6.21}$$

Again $\overline{\mathbf{k}}$ is the 6k × 1 vector of gyroscopic and centrifugal forces as well as Coriolis forces. Finally, the applied forces and moments and the reaction forces and moments are located in the 6k × 1 vectors $\overline{\mathbf{p}}^e$ and $\overline{\mathbf{p}}^r$:

$$\overline{\mathbf{p}}^e = [\mathbf{F_1^{eT}}|\cdots|\mathbf{F_k^{eT}}|\mathbf{M_1^{eT}}|\cdots|\mathbf{M_k^{eT}}]^T \tag{6.22}$$

$$\overline{\mathbf{p}}^r = [\mathbf{F_1^{rT}}|\cdots|\mathbf{F_k^{rT}}|\mathbf{M_1^{rT}}|\cdots|\mathbf{M_k^{rT}}]^T \tag{6.23}$$

Finally, we multiply the equation system (6.19) from the left with the transposed, global Jacobi matrix $\overline{\mathbf{J}}^T$, so that it is formulated completely in generalised coordinates. This yields equilibrium of forces in matrix form:

$$\mathbf{M\ddot{q}} + \mathbf{k} = \mathbf{Q} \tag{6.24}$$

The product $\overline{\mathbf{J}}^T\overline{\mathbf{M}}\overline{\mathbf{J}}$ yields the mass matrix $\mathbf{M}$. Similarly, $\overline{\mathbf{J}}^T\overline{\mathbf{k}}$ yields $\mathbf{k}$, the vector of the generalised gyroscopic forces, and $\overline{\mathbf{J}}^T\mathbf{F}^e$ yields the vector of the generalised forces $\mathbf{Q}$. Here $\mathbf{M}$ is dependent upon the generalised coordinates $\mathbf{q}$ and t, and $\mathbf{k}$ and $\mathbf{Q}$ are dependent upon $\mathbf{q}$, $\dot{\mathbf{q}}$ and t. Last but not least, we should note that the reaction forces are dispensed with as a result of the multiplication by $\overline{\mathbf{J}}^T$. We therefore have a system of ordinary differential equations to solve because the algebraic equations of the constraints have disappeared with the reaction forces.

Lagrange approach

The focus of the Newton–Euler approach described in the previous section was the drawing up of Newton and Euler equations for each body and the conversion of the resulting overall system into generalised coordinates so that the constraint equations are dispensed with. The Lagrange approach takes a different and particularly elegant route. It starts from the premise that the generalised inertial forces and the generalised applied forces cancel each other out. For the formulation of the generalised inertial forces Q_i^I we require the total kinetic energy T of the system, which of course must also be formulated in the form of generalised coordinates:

$$Q_i^I = \frac{\partial T}{\partial q_i} - \frac{d}{dt}\left(\frac{\partial T}{\partial \dot{q}_i}\right) \tag{6.25}$$

The first subterm represents the generalised inertial forces that arise as a result of the change of position of the system, thus, for example, the Coriolis force. Opposing this part is the second component, which describes the rate of change of the generalised impulses. The above-mentioned premise

$$Q_i^I = -Q_i \qquad (6.26)$$

yields the Lagrange equation

$$\frac{d}{dt}\left(\frac{\partial T}{\partial \dot{q}_i}\right) - \frac{\partial T}{\partial q_i} = Q_i \qquad (i = 1, 2, \dots, n) \qquad (6.27)$$

By drawing up a formula for kinetic energy and the conversion of applied forces into their generalised form we can obtain the equations of motion directly by substituting into equation (6.27). This is particularly simple because kinetic energy is a scalar that contains no higher derivatives with respect to time than the velocities. These are significantly easier to determine than the accelerations.

Formulation in hardware description languages

The primary purpose of analogue hardware description languages is for the modelling of analogue electronic components for a circuit simulator. The variables considered in this application — voltage and current — correspond with the duality of a potential and a flow and can be represented by other quantities in accordance with the analogies described in Section 3.2.2. Although the text-based formulation of the mechanical model is based upon accelerations, velocities, positions, and forces, the underlying calculations take place in accordance with the analogies of the potentials and flows available.

Furthermore, the preceding section has shown that the selection of the considered unknowns of multi-body mechanics is attributed decisive importance. In electronics the unknowns are normally in the form of node voltages, which is because of the nodal analysis that is prevalent in circuit simulation. In the system-oriented modelling of mechanics, on the other hand, it is of decisive importance to specify a suitable set of generalised coordinates. For holonomous systems, which can be described using generalised coordinates, the — sometimes very complex — constraint equations are dispensed with. As was shown by the relatively simple example from Figure 6.2, it is not a question of selecting from a fund of existing coordinates, but one of an independent engineering task.

The methods described supply sets of ordinary differential equations in symbolic form. These can easily be formulated in analogue hardware description languages. This is true under the prerequisite that the size of the equation set remains within limits. In Section 7.2.3 the obtaining and formatting of the equations of motion for an automotive wheel suspension system using the Lagrange approach is illustrated.

6.2.3 Object-oriented modelling

Introduction

The use of generalised coordinates in object-oriented modelling raises two problems. Firstly, it poses the question of how we should determine the generalised coordinates from the very limited perspective of an element. Secondly, the local Jacobi matrices, which describe how the local coordinates arise from the totality of the generalised coordinates, have to be set up. Both questions necessitate the global perspective of mechanics, i.e. the local consideration that has brought so many benefits in electronics is lost. In other words: When generalised coordinates are used the consideration of a multibody system generally results in the completion of the drawing up of the Newton–Euler equation system

$$\mathbf{M}\ddot{\mathbf{q}} + \mathbf{k} = \mathbf{Q} \tag{6.28}$$

using hardware description languages, based upon models for rigid bodies, springs, dampers etc. A more promising approach would seem to be to add the automated creation of symbolic equations of motion by a suitable programme and thus select system-oriented modelling.

Object-oriented modelling thus cannot be performed directly using generalised coordinates. However, if we free ourselves from the generalised coordinates and in particular permit a greater number of unknowns, then the question is reformulated. The work of Suescun *et al.* [391] provides a first approach to the modelling of multidimensional mechanics in hardware description languages (VHDL-AMS). Here the position of the body is given in natural coordinates, which occur in two forms: Firstly, they are given as cartesian coordinates for certain points on the body. These marked points may be contact points of joints, springs and dampers. Secondly, unit vectors are introduced as natural coordinates, in order to specify axes of rotation. According to Suescun *et al.* the mass matrix $\mathbf{M}$ of a body is constant if a sufficient quantity of natural coordinates are considered. This represents the vector $\mathbf{q}$ of the natural coordinates on the inertial force $\mathbf{Q}_I$ (with respect to the natural coordinates):

$$\mathbf{Q}_I = -\mathbf{M}\ddot{\mathbf{q}} \tag{6.29}$$

The natural coordinates are modelled in the hardware description languages as potentials (across), the forces and moments as flows (through). In addition there are algebraic constraint equations in quadratic (for planar mechanics) and cubic (for 3D mechanics) form, which hold constant the constellation of points in relation to each other and the length of the unit vectors. In addition there is a VHDL-AMS module for the gravitation that is suspended on the rigid body model. Also present are models for joints, springs and dampers. The models mentioned are put together using a circuit editor. The corresponding system of DAE is then solved in a circuit simulator. Finally, we should also mention that no simulation results are shown in [391].

In addition to the general principle there is one special case, for which the development of arbitrarily connectable models has been known for a long time. The prerequisite for this is that the movement in the system only takes place in one translational or rotational dimension, or that the movements in the system can be broken down into one-dimensional movements that are independent of each other. Then the generalised coordinates coincide with the cartesian coordinates or the angular coordinates, the Jacobi matrices are trivial and the mechanical forces/moments and velocities/angular velocities can be represented directly by potentials and flows. Applications are any pure translational movements and one-dimensional rotational movements, such as in a drive train with motor, gearbox and mechanical load. In the following suitable models for the basic elements mass, spring, damper, power source and position source[1] will be described.

Basic model

The basic elements mass, spring, and damper can be formulated for both translational and rotational movements. In the following the translational version will be given, with positions (instead of the velocities) being used as potentials. It is formulated in the hardware description language VHDL-AMS. We first begin with the model of mass inertia for translational movements inertia_trans, see Hardware description 6.1. This model follows the equation

$$F = -(m \cdot \ddot{x}) - (m \cdot g) \qquad (6.30)$$

and thus describes both the inertia and also the acceleration due to gravity. If the acceleration due to gravity does not lie in the direction of the translation, the parameter GRAVITY, i.e. g, is set to 0. Otherwise the model follows the convention that an acceleration in the direction of x gives rise to negative forces and vice versa.

```
LIBRARY disciplines;              -- Reference to a package with the
USE disciplines.Kinematic_system.all; -- mechanics declarations
ENTITY inertia_trans IS                 -- Interface description
   GENERIC (m, g: REAL);                      -- Mass, gravity
   PORT    (TERMINAL p, n: kinematic);          -- Terminals
end inertia_trans;
ARCHITECTURE simple OF inertia_trans IS        -- Architecture
-- Declaration of potential/flow quantity=deflection x/force F ...
  QUANTITY tdisp ACROSS tforce THROUGH p TO n;
  BEGIN
     tforce == -(m*tdisp'DOT'DOT) - (m*g);       -- Basic equation
  END simple;
```

Hardware description 6.1   Model of a mass for translational movement in VHDL-AMS

---

[1] Like a voltage or current source, but supplying a position.

Now to spring and damper models. For the spring model the applied force is dependent upon the position, i.e. upon the distortion of the spring. The damping force, on the other hand, is proportional to the relative velocity of the two terminals of the damper model, and thus primarily describes the Stokes' friction of a viscous fluid, such as for example in an automotive shock absorber. The following equations form the basis:

$$F_{spring} = -k(x_p - x_n - l_0)$$
$$F_{damper} = -b(v_p - v_n) \tag{6.31}$$

The appropriate conversion is found in Hardware descriptions 6.2 and 6.3.

```
LIBRARY disciplines;            -- Reference to a package with the
USE disciplines.Kinematic_system.all;-- mechanics declarations
ENTITY spring_trans IS                  -- Interface description
  GENERIC (k, 10: REAL);-- Spring constant, basic spring length
  PORT (TERMINAL p, n: kinematic);                 -- Terminals
end spring_trans;
ARCHITECTURE simple OF spring_trans IS          -- Architecture
        -- Declaration of potential/flow = deflection/force ...
  QUANTITY tdisp ACROSS tforce THROUGH p TO n;
BEGIN
  tforce == -k * (tdisp - 10);           -- Basic equation
END simple;
```

Hardware description 6.2   Spring model for translational movements

In both cases the spring or the damping force is first calculated and correspondingly applied. This force is applied in the negative direction. For the spring this is consistent with the convention that positive forces increase the current positional value. The spring force at terminal p is oriented such that the spring length tends towards the equilibrium l_0. At terminal n the force is correspondingly oriented in the opposite direction. For the damper, the convention applies that positive forces increase the relative distance of the two position terminals. The damping force resists a positive, relative velocity. The descriptions for the application of forces and velocities will not be illustrated here. They correspond with the applicable descriptions of sources for currents and voltages.

```
LIBRARY disciplines            -- Reference to a package with the
USE disciplines.Kinematic_system.all; -- mechanics declarations
ENTITY damper_trans IS                 -- Interface description
  GENERIC (b: REAL);                         -- Damper constant
  PORT (TERMINAL p, n: kinematic);                 -- Terminals
end damper_trans;
ARCHITECTURE simple OF damper_trans IS -- Architecture 'simple'
        -- Declaration of potential/flow = deflection/force ...
  QUANTITY tdisp ACROSS tforce THROUGH p TO n;
```

```
BEGIN
    tforce == -b * tdisp'DOT;                    -- Basic equation
END simple;
```

Hardware description 6.3 Damper model for translational movements

6.2.4 Example: wheel suspension

Starting from these basic models we can now put together more complex models. A wheel suspension will serve as an example. Let us first of all set up the framework for the consideration of a modelling process. We assume that only the vertical movement of the wheel and the vehicle body is to be considered. Furthermore, the condition is imposed that the centre of gravity of the vehicle is located mainly in the centre of the vehicle and thus the axles are uniformly loaded. In this case the movements of the axles are almost independent of each other, which means that we can restrict ourselves to the consideration of one axle. If we further assume that the road conditions are the same for the left and right-hand wheel, then for reasons of symmetry it is completely adequate to consider only one wheel including half an axle and a quarter of the car body. Using the assumptions described yields a two-mass oscillator, which describes the vertical dynamics very well, see Figure 6.3.

The mass m_a describes the wheel and the associated part of the axle and m_b describes a quarter of the body. Both masses are of course subject to gravity, but also to the forces that are exerted by the adjacent springs and dampers. Shock absorbers and body springs themselves are characterised by the parameters b and k_s respectively. The tyres can also be considered as springs, but with a spring constant k_w that lies around an order of magnitude above that of the body spring. The

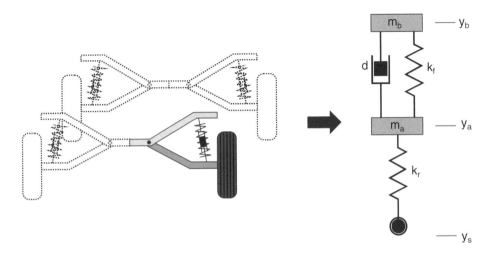

Figure 6.3 Modelling of a wheel suspension by a two-mass oscillator

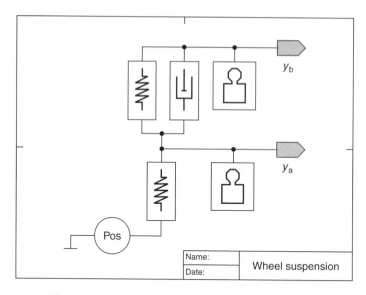

Figure 6.4   Schematic diagram of a wheel suspension

Table 6.1   Parameters

Parameter	Value
m_a	50 kg
m_b	250 kg
k_s	25 500 N/m
k_w	250 000 N/m
B	2000 Ns/m
$l_{0,\text{Spring}}$	0.2 m
$l_{0,\text{Tyre}}$	0.03 m
G	9.81 m/s^2

damping effect of the tyres can be disregarded here. The system is one-dimensional because only the vertical movement of the masses is being considered. It has two degrees of freedom, the y-positions of the two masses. The y-position of the road serves as the stimulation. Driving over a step of a few centimetres is modelled by imposing a jump of corresponding height. This system can be assembled directly from the basic elements developed above in the form of a schematic diagram, see Figure 6.4.

After suitable parameterisation, simulation can take place without further modelling expense. The parameters in Table 6.1 are used for the simulation shown in Figure 6.5.

The situation considered in the simulation corresponds with driving over a 5 cm high step at a right angle, i.e. the left-hand and right-hand wheel experience the same deflection. Thus the symmetry condition is fulfilled. At the beginning of the journey the spring forces of springs and tyres correspond with the respective weight

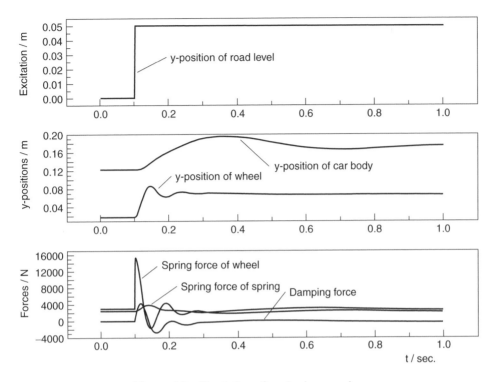

Figure 6.5 Simulation of a wheel suspension

and ensure that body and wheel dip in relation to gravity. Whilst the road level y_s rises suddenly by 5 cm, the tyre springs are correspondingly compressed and quickly build up a force of approximately 15 000 N. As a result, the wheel is pushed upwards, the tyre spring relaxes and the force in question eases. However, the body spring is compressed by the movement of the wheel and corresponding forces are transmitted to the body. In accordance with the mass and spring constants, vibrations are observed in the range of around one hertz. In the case of the wheel the vibrations lie in the range of around ten hertz. Thus the body requires significantly longer to take on its new y-position. Finally, it should be noted that the damping force works to counter the relative movement between wheel and body and thus allows the vibrations to decay. The simulation requires few CPU seconds on a SUN Sparc 20 workstation.

6.2.5 Further applications

Introduction

In the following a few other applications will be presented as examples, thereby illustrating the possibilities of multibody modelling using hardware description languages. The representation takes into account both mechatronic and micromechatronic systems.

Mechatronics

In [254] Makki *et al.* describe an electronically controlled window winder mechanism for cars. On the mechanical side a direct current motor, a gearbox, a rack for the conversion of the rotational motion into a translational movement, a mechanical load — the window pane — and a mechanical stop are envisaged. In addition to this there is a force sensor that allows the drive to be switched off in the event of large counterforces. This typically corresponds with a situation in which objects are trapped by the window-pane whilst the window is raised. In this case movement is restricted to a rotary — or after the rack a translational — dimension. For this reason the system described can be simply assembled from basic models, each of which corresponds with one of the named components.

Other examples can be found in Donnelly *et al.* [84], who describe an electronically controlled hydraulic braking system, or in Mikkola [269], who uses hardware description languages to model and simulate diesel-electric ship drives.

Micromechatronics

For the class of so-called 'suspended' MEMS, Mukherjee and Fedder [282] have developed an approach based upon multibody mechanics. Classical applications for this approach are, for example, seismically suspended masses of acceleration sensors and resonators, see Figure 6.6. The structure of interest is broken down into individual parts such as springs, masses, dampers, etc., for which models are available. Thus a micromechanical model can be assembled from the basic models. This strategy is very well suited to the approach that is also selected here of formulation in hardware description languages, because these continuously support the hierarchical structure of models.

In the NODAS system in [103], Fedder and Jing go beyond multibody systems made up of rigid bodies by including elastic components on the basis of hardware description languages. The following components have been implemented in NODAS as described in [103]: A bending beam, a rigid plate, an electrostatic

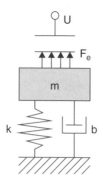

Figure 6.6   Electrically excited resonator in the form of a multibody system

comb actuator, and an anchor (which corresponds with a fixed suspension). Implementation first raises the question of differentiating between a global and local coordinate system. Initially all considerations of an element are local. However, the element can also be given global coordinates, which can be used to solve a calculation of the operating point. Thus the correct values of the global coordinates are set automatically, whereas the actual calculations generally take place using a further set of variables that only give values relative to the operating point.

The model of the bending beam was developed on the basis of a mechanical structural analysis. The equation for the beam takes the form:

$$\mathbf{F_{beam}} = \mathbf{M\ddot{u}} + \mathbf{B\dot{u}} + \mathbf{Ku} \tag{6.32}$$

where $\mathbf{F_{beam}}$ represents the vector of forces and moments at the beam, $\mathbf{u}$ represents the vector of the translational and rotational degrees of freedom of the beam, $\mathbf{M}$ represents the mass matrix, $\mathbf{B}$ the damping matrix and $\mathbf{K}$ the stiffness matrix. In principle this follows the beams presented in [34] and in Section 6.3.2, although NODAS is more interested in the global movement and not in the deformation of a continuum. Furthermore, the local mass, damping and stiffness matrices are formulated directly in the hardware description language, which may cause these symbolic equations to explode in the event of more complex elements.

However, in many cases physical modelling, as used in the previous examples, is not possible or would be associated with great expense. In such cases it is often worthwhile to move to experimental modelling. In this approach an experiment does not necessarily consist of measurements on a real system, but often consists of field and continuum simulations, for example based upon finite elements. In this manner, the simulation can be run in advance of manufacture by the use of experimental models. Pure table models, such as for example in Romanowicz *et al.* [350] or Swart *et al.* [394], are an example. However, these table models, with their data list of identification pairs, can be represented by compression into relatively simple equations. This is shown by Teegarten *et al.* [397], who also supply a lovely example of the mixing of physical and experimental modelling based upon a micromechanical gyroscope.

6.3 Continuum Mechanics

6.3.1 Introduction

The previous section dealt with multibody mechanics, the main characteristic of which is the consideration of a collection of bodies connected together by joints and suspensions. The validity of this abstraction depends upon the formulation of the question. In particular, the bending of mechanical components is often not an undesirable side-effect, but is essential to the functioning of the system. Now, if the form of bending plays a significant role in the system behaviour then we cannot avoid the consideration of the continuum in the modelling. The associated mechanics, and in particular its representation in hardware description languages, are the subject of this section.

We can initially differentiate between whether the consideration is to be performed statically or dynamically. For the static case each mechanical position may be assigned an electrical quantity. Here only the steady state is considered. In the dynamic case velocities and accelerations of mechanical quantities also play a role, so that phenomena such as mechanical resonance are also considered. A further distinction is supplied by the selection of a desired level of abstraction. It is a fundamental truth of continuum mechanics that elasticity and mass spread continuously, thus giving rise to an infinite number of degrees of freedom. As is described in more detail in what follows, we can perform the modelling of mechanical continua on the basis of (geometric) structure, physical equations, and experimental data. In this context the reader is referred to a corresponding classification of modelling approaches in Section 2.4.

6.3.2   Structural modelling

Introduction

Structural modelling traces the generation of a model back to the composition of basic models. In the case of continuum mechanics these basic models may be finite elements, for example. Due to the generality and high degree of adoption of finite elements in the framework of structural modelling, we will deal exclusively with this approach. Bathe [19], Gasch and Knothe [113] and Knothe and Wessels [202] supply a good overview of the methods of finite elements in their works. For the modelling of finite elements, as in the Ritz procedure (see within Section 6.3.3), we work on the basis of interpolation functions. However, these are not formulated globally for the whole structure here, but locally for the finite element. Thus the main difficulties of the Ritz procedure are removed. If the models of the finite elements are available, modelling is a purely geometric task, which primarily represents a breakdown of the continuum. In this context we also speak of a meshing, in which finer resolutions buy more precision at the expense of greater simulation time.

Up until now, finite elements have typically only been used to investigate the component level, disregarding the system context. The following sections will show that finite elements can be drawn into a circuit simulation on the basis of hardware description languages. As we will show in what follows, the differential equation solver of the circuit simulator in question is thus entrusted with the calculation of the equations of the finite elements. The dynamic coupling of electronics and mechanics then takes place automatically. Overall, this opens up a simpler, faster and more secure way of modelling mechanical continua that is compatible with hardware description languages and thus also with circuit simulation.

Finite elements

In principle, finite elements can be used in many fields of engineering science. Our discussion is based upon the field of structural mechanics. Thus the following quantities have to be linked together: displacements, forces, strain, and applied loads, which act as a trigger here. Depending upon the application, different finite elements are used, which vary in structure, number of nodes and degrees of freedom. Figure 6.7 shows a selection of finite elements of structural mechanics.

The degrees of freedom of the finite elements can be of both a translational (u_x, u_y, u_z) and a rotational (r_x, r_y, r_z) nature. The numerical complexity of the calculation increases with their number. Fundamentally, the element selected should fulfil the question formulated with as few superfluous degrees of freedom as possible. In addition, symmetry considerations are used to keep the number of finite elements as low as possible.

In the following, an approach will be presented that allows the finite elements of structural mechanics to be represented in hardware description languages. This is based upon the work of Pelz *et al.* [333] and Bielefeld *et al.* [33] and [34]. Information on the mechanical foundation can be found in [19], [113] or [202]. We should mention at this point that Haase *et al.* [131] to some degree follow a similar approach in a subsequent work by linking system matrices that originate from a commercial FE simulator into a circuit simulator.

The formulation of the finite elements firstly requires that a mass matrix, a stiffness matrix, and a load vector are generated for each element. As in many other works the damping matrix is initially disregarded. Secondly, in a FE simulator, these element matrices are combined into a global equation system, according to

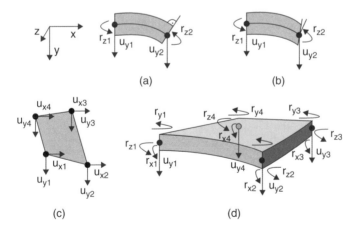

Figure 6.7 Selection of finite elements from structural mechanics: (a) Shear-resistant 1-D beam, two nodes, two degrees of freedom per node (u_y, r_z) (b) Non shear-resistant 1-D, two nodes, two degrees of freedom per node (u_y, r_z) (c) Plane element, four nodes, two degrees of freedom per nodes (u_x, u_y) (d) Plate element, four nodes, four degrees of freedom per node (u_y, r_x, r_y, r_z)

the structure of the mechanics. This must either be completed during the modelling or in the circuit simulator.

A sensible starting point in the drawing up of the element matrices is the principle of virtual displacement. A virtual displacement is a small displacement superimposed upon the actual displacement, which fulfils the geometric boundary conditions and otherwise brings about no gaps or overlapping of the continuum.

The principle of virtual displacements demands that the virtual displacement energy is equal to the virtual work of the external forces for each permitted virtual displacement. This yields the basic equation that is drawn up for the whole continuum. Now the components of the individual elements in the basic equation should be taken into account. This would require knowledge of the continuous displacements over the entire element. However, because we want to operate using only the displacements of the nodes of the finite elements, it is necessary to approximate the continuous displacements from the node displacements. This is done with the aid of interpolation functions that are often created in the form of polynomials. Thus the continuous displacements are approximated from the node displacements, and using the displacement/strain relationship these are transformed into the strains of the element. Using the underlying law of matter we find the stresses from the strains. Using the quantities determined in this manner, the strain energy can be integrated over the element range and summed over all elements. The integration is significantly simplified by the use of interpolation functions, which — as noted before — typically are polynomial.

By contrast, the virtual work of the external forces is based upon the excitation forces, stresses at the edge of the body, and body forces such as weight. The associated proportions of (virtual) work are again calculated from the node displacements by integration over the range in question and summed for all elements. Finally, the total virtual strain energy is equated to the total virtual work of the external forces. In the static case this yields the following equation system:

$$\mathbf{Ku} = \mathbf{p} \qquad\qquad (6.33)$$

where $\mathbf{K}$ represents the system stiffness matrix, $\mathbf{u}$ the node displacements, and $\mathbf{p}$ the converted body and contact forces at the nodes. The system stiffness matrix is found from the suitable addition of the element stiffness matrices. In the kinetic case there are also inertia forces and the equation is formulated as follows:

$$\mathbf{M\ddot{u} + Ku} = \mathbf{p} \qquad\qquad (6.34)$$

where $\mathbf{M}$ represents the system mass matrix, which, in a similar way to the system stiffness matrix, is found by a suitable summing of the element mass matrices. The system mass matrix is linked with the accelerations of the displacements. In this discussion both equation systems correspond with the equilibrium principle.

If we want to represent finite elements in hardware description languages, then it initially appears logical to first draw up the differential equation system resulting from the collection of finite elements in symbolic form, and then to directly formulate this in a hardware description language. In theory this is correct. However, the handling of the equations causes massive problems. This is firstly the case if we want to parameterise the elements geometrically and not on the basis of the entries in the element matrix. The same applies in the nonlinear consideration if the mass and stiffness matrices of the finite elements are dependent upon the current state of deformation and have to be drawn up afresh depending upon deflection. In both cases the complete rule for the creation of the element matrices must be included in the equation system, as must the conversion from the element matrices into the system matrix. This allows the volume of equations to explode and the resulting equation system is thus beyond any meaningful calculation.

It therefore makes sense to initially consider the finite elements individually and to build the rule for the creation of the element matrices into the model in question. This could, for example, be achieved by embedding a C routine, capable of generating suitable element matrices as required, into the model. This corresponds with the numerical simulation of multibody systems. The question is also raised of how to move from element behaviour to system behaviour. Ideally, the system behaviour would be found by composing the finite elements in a circuit simulator. This first requires a link between electronic quantities and mechanical degrees of freedom. Here mechanical deflections are represented by electrical potentials and mechanical forces and moments by electrical currents. The linking of two finite elements effects a scleronomic[2] constraint between the element degrees of freedom in question, and thus the amalgamation of the degrees of freedom in question to a single system degree of freedom. This is the expected behaviour for voltages and positions. Furthermore, the currents are added at these points, as is also expected of forces and moments.

The generation of element matrices

In what follows the example of the shear-resistant beam element will be used to demonstrate how such a finite element can be formulated in hardware description languages. To achieve this two main problems have to be solved. Firstly the mass and stiffness matrices in question have to be generated. Secondly the element matrices have to be transformed into the system matrices, which represents the behaviour of the entire structure.

The beam element is shown in Figure 6.8 and has two nodes, k and l, each with two degrees of freedom, the deflection in the y-direction u_y and the rotation about the z-axis r_z. Both the shear deflection in the x-direction and the structural damping are disregarded in this model.

The stiffness $B_i = EI_{zz}$ and the mass distribution $\mu_i = \rho A_i$ are assumed to be constant over the length of the beam, where E is the modulus of elasticity, I_{zz} the moment of inertia, ρ the density of the beam material and A_i the cross-section of

---

[2] Scleronomic constraints are not changeable.

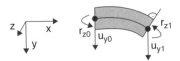

Figure 6.8 Degrees of freedom of the shear-resistant beam element at the nodes k and l: u_y (deflection in y-direction), r_z (rotation about the z-axis)

the beam section i. The beam load is concentrated by p_{i0} and p_{i1} on the nodes 0 and 1 of the beam element. In the shear-resistant case and for small deflections the stiffness matrix $\mathbf{K}_i$, the mass matrix $\mathbf{M}_i$ and the load vector $\mathbf{p}_i$ of the ith beam element are independent of the deflection. If we select the interpolation functions $h_1 \ldots h_4$ in the variables ξ for the approximation of the continuous displacements as follows:

$$h_1(\xi) = 1 - 3\xi^2 + 2\xi^3$$
$$h_2(\xi) = -\xi(1-\xi)^2 l_i$$
$$h_3(\xi) = 3\xi^2 - 2\xi^3$$
$$h_4(\xi) = \xi^2(1-\xi)l_i$$

(6.35)

then we find the following element matrices and vectors, see Gasch and Knothe [113]:

$$
\mathbf{K}_i = \frac{B_i}{l_i^3}
\begin{bmatrix}
12 & -6l_i & -12 & -6l_i \\
-6l & 4l_i^2 & 6l_i & 2l_i^2 \\
-12 & 6l_i & 12 & 6l_i \\
-6l & 2l_i^2 & 6l_i & 4l_i^2
\end{bmatrix}
$$

$$
\mathbf{M}_i = \frac{\mu_i l_i}{420}
\begin{bmatrix}
156 & -22l_i & 54 & 13l_i \\
-22l_i & 4l_i^2 & -13l_i & -3l_i^2 \\
54 & -13l_i & 156 & 22l_i \\
13l_i & -3l_i^2 & 22l_i & 4l_i^2
\end{bmatrix}
$$

(6.36)

$$
\mathbf{p}_i = p_{i0}l_i
\begin{bmatrix}
7/20 \\
-l_i/20 \\
3/20 \\
l_i/3
\end{bmatrix}
+ p_{i1}l_i
\begin{bmatrix}
3/20 \\
-l_i/30 \\
7/20 \\
l_i/20
\end{bmatrix}
$$

The equation system for such an element thus takes the form:

$$\mathbf{M}_i \ddot{\mathbf{u}}_i + \mathbf{K}_i \mathbf{u}_i = \mathbf{p}_i \qquad (6.37)$$

where

$$\mathbf{u}_i = [u_{y0}, r_{z0}, u_{y1}, r_{z1}]^T$$

where $\mathbf{u}_i$ represents the element displacement vector, and thus the degrees of freedom.

Now, if the behaviour of a mechanical continuum is to be reconstructed in a circuit simulator it is reasonable to keep the modelling close to the actual determination of the simulator. In our case this means that the mechanics model is formulated 'electronically'. For this purpose a network of capacitors, inductors and current sources is drawn up, see Figure 6.9. If we consider the associated admittance matrix we notice that just like the mass and stiffness matrices it is symmetrical and its leading diagonal consists of positive entries.

The task now is to find an LC network, the admittance matrix of which coincides with the mass and stiffness matrix of the mechanics. To a certain degree this corresponds with the drawing up of a type of equivalent circuit. However, we will see later that the formulation in hardware description languages does not rest upon components, but uses the underlying equations. Let us first consider the circuit in Figure 6.9 and draw up Kirchhoff's current law for the four nodes, i.e. four degrees of freedom:

$$\sum_{j=1}^{4} (i_{ij,L} + i_{ij,C}) = i_i \qquad i = 1\ldots 4, \qquad (i_{ii,L} = 0, \quad i_{ii,C} = 0) \tag{6.38}$$

Using the current–voltage relationships this yields the following equations:

$$\sum_{j=1}^{4} \left(\frac{1}{L_{ij}} \int u_{ij} \, dt + C_{ij} \dot{u}_{ij} \right) = i_i \qquad i = 1\ldots 4, \quad L_{ij} = L_{ji}, \quad C_{ij} = C_{ji} \tag{6.39}$$

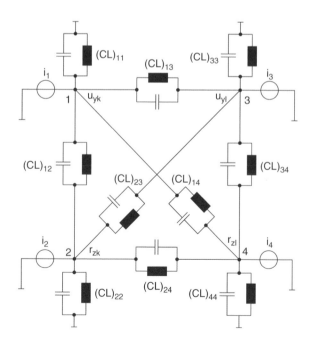

Figure 6.9 LC network with current sources for the modelling of finite beam elements

These four equations are differentiated once with respect to time and then rearranged to give:

$$\sum_{j=1}^{4} C_{ij} \ddot{u}_{ij} + \sum_{j=1}^{4} \frac{1}{L_{ij}} u_{ij} = \dot{i}_i \qquad i = 1 \ldots 4 \qquad (6.40)$$

The four degrees of freedom of the beam element u_{yk}, u_{yl}, r_{zk} and r_{zl} should now be represented by the potentials φ_1, φ_2, φ_3 and φ_4, which for this reason are used here for the branch voltages u_{ij}. The following is true:

$$u_{ij} = \varphi_i - \varphi_j \qquad (i \neq j)$$

$$u_{ii} = \varphi_i \qquad (6.41)$$

Substituting into the above formula yields:

$$C_{ii} \ddot{\varphi}_i + \sum_{i \neq j} C_{ij} (\ddot{\varphi}_i - \ddot{\varphi}_j) + \frac{1}{L_{ii}} \varphi_i + \sum_{i \neq j} \frac{1}{L_{ij}} (\varphi_i - \varphi_j) = \dot{i}_i \qquad i = 1 \ldots 4 \quad (6.42)$$

After rearranging this yields in vector notation:

$$\mathbf{C} \ddot{\varphi} + \mathbf{L} \varphi = \mathbf{i} \qquad (6.43)$$

where:

$$\ddot{\boldsymbol{\varphi}} = [\ddot{\varphi}_1, \ddot{\varphi}_2, \ddot{\varphi}_3, \ddot{\varphi}_4]^T$$

$$\boldsymbol{\varphi} = [\varphi_1, \varphi_2, \varphi_3, \varphi_4]^T$$

$$\dot{\mathbf{i}} = [i_1, i_2, i_3, i_4]^T$$

$$\mathbf{C} = \begin{bmatrix} C_{11} + C_{12} + C_{13} + C_{14} & -C_{12} & -C_{13} & -C_{14} \\ -C_{12} & C_{12} + C_{22} + C_{23} + C_{24} & -C_{23} & -C_{24} \\ -C_{13} & -C_{23} & C_{13} + C_{23} + C_{33} + C_{34} & -C_{34} \\ -C_{14} & -C_{24} & -C_{34} & C_{14} + C_{24} + C_{34} + C_{44} \end{bmatrix}$$

$$\mathbf{L} = \begin{bmatrix} \frac{1}{L_{11}} + \frac{1}{L_{12}} + \frac{1}{L_{13}} + \frac{1}{L_{14}} & -\frac{1}{L_{12}} & -\frac{1}{L_{13}} & -\frac{1}{L_{14}} \\ -\frac{1}{L_{12}} & \frac{1}{L_{12}} + \frac{1}{L_{22}} + \frac{1}{L_{23}} + \frac{1}{L_{24}} & -\frac{1}{L_{23}} & -\frac{1}{L_{24}} \\ -\frac{1}{L_{13}} & -\frac{1}{L_{23}} & \frac{1}{L_{13}} + \frac{1}{L_{23}} + \frac{1}{L_{33}} + \frac{1}{L_{34}} & -\frac{1}{L_{34}} \\ -\frac{1}{L_{14}} & -\frac{1}{L_{24}} & -\frac{1}{L_{34}} & \frac{1}{L_{14}} + \frac{1}{L_{24}} + \frac{1}{L_{34}} + \frac{1}{L_{44}} \end{bmatrix}$$

Let us now return to the equations of the ith mechanical, finite beam element:

$$\mathbf{M_i \ddot{u}_i + K_i u_i = p_i} \qquad (6.37)$$

where

$$\mathbf{u}_i = [u_{y0}, r_{z0}, u_{y1}, r_{z1}]^T$$

This equation system has the same structure as the LC network, see equation (6.43). We now have to identify the individual components of the two matrix equations with each other, i.e.:

$$\ddot{\mathbf{u}}_i \mathrel{\hat{=}} \ddot{\varphi}$$

$$\mathbf{u}_i \mathrel{\hat{=}} \varphi$$

$$\mathbf{M}_i \mathrel{\hat{=}} \mathbf{C} \tag{6.44}$$

$$\mathbf{K}_i \mathrel{\hat{=}} \mathbf{L}$$

$$\mathbf{p}_i \mathrel{\hat{=}} \mathbf{i}$$

The degrees of freedom of the finite beam elements are directly represented by the potentials, i.e. the node voltages. The same applies for the associated accelerations.

In order to balance the matrix entries in question, the negative entries of the mass matrix m_{ij} are used for the capacitance entries in the secondary diagonals, the sum of the involved mass coefficients are used in the leading diagonal:

$$C_{ij} = -m_{ij} \qquad (i \neq j)$$
$$C_{ii} = m_{ii} + \sum m_{ij} \quad (i \neq j) \tag{6.45}$$

In a similar way, the entries for the inductance matrix are formed from the stiffness coefficients k_{ij}:

$$L_{ij} = \frac{1}{k_{ij}} \qquad (i \neq j)$$
$$L_{ii} = \frac{1}{k_{ii} + \sum k_{ij}} \qquad (i \neq j) \tag{6.46}$$

The equations (6.45) and (6.46) ensure that the matrices $\mathbf{M}_i$ and $\mathbf{K}_i$ described by $\mathbf{C}$ and $\mathbf{L}$ are represented with sufficient precision, i.e. there is a good correspondence between equation systems (6.37) and (6.43). Correction terms obtained from the summing term are also added into the leading diagonals of $\mathbf{C}$ and $\mathbf{L}$. These ensure that the LC circuit yielded from the matrices satisfies Kirchhoff's laws and, in particular, that the currents linked by the nodes add up to zero. This corresponds with a variation of the LC branches from the nodes 1 to 4 to the mass, which thus characterises not the relationships between every two degrees of freedom, but only the relationship of the degree of freedom to ground.

Finally, the derivative of the currents $\mathbf{i}$ are derived as follows. The loads of the beam element concentrated at the nodes p_{i0} and p_{i1} are converted by equation (6.36) into the element load vector. The components of this are then integrated and, in the form of current, put into the nodes of the associated degree of freedom. This takes place for every time step, so that time-variant loads can also be taken into account.

The finite elements are formulated in the analogue hardware description language MAST of the Saber circuit simulator and this formulation is primarily based upon introducing two current sources between the nodes i and j for an LC branch, which satisfy the following equations:

$$i_{ij,C} = m_{ij} \dot{u}_{ij} \qquad \text{where } m_{ij} \text{ is from } \mathbf{M}$$

$$i_{ij,L} = k_{ij} \int u_{ij}\, dt \qquad \text{where } k_{ij} \text{ is from } \mathbf{K} \tag{6.47}$$

In addition there are two further current sources for each degree of freedom, which represent the connection to the ground and — as demonstrated above — the external excitations p_{i0} and p_{i1} at each beam element.

Composition of the system matrix

In the previous section an element matrix was put together for the beam element, the four degrees of freedom of which are represented by the potentials at the four terminals of the element. The currents at the nodes in question describe the integral of the associated forces and moments, depending upon whether the degree of freedom is a translational or rotational deflection. In particular, the components of the exciting forces and moments that are assigned to the elements adjoining the nodes are also added to the currents at a node. Thus it is not necessary to explicitly draw up the system matrix. Its solution is found implicitly from the interconnection of the finite elements.

Example: beam with various boundary conditions

Two examples will be considered to illustrate the element model described above, a cantilever beam with and without an additional support point, see Figure 6.10. The second case, in particular, cannot simply be mastered by either analytical equations or finite differences. The beam of length l itself is modelled by 40 finite beam elements. The excitation consists of the pulsed force F_y, which is applied to the beam for eight seconds and then removed again. The outputs are the deflections in the y-direction at $x = 0.25l, 0.5l, 0.75l$ and $1.0l$, see Figure 6.11. In the first case there is an oscillation, the amplitude of which is more strongly marked towards the end of the beam, and which is in phase at each point. The additional support in the second case fundamentally alters the behaviour of the beam. Firstly, the natural frequency of the system increases, secondly the node moves downwards at $x = 0.25l$ due to the lever effect of the free end of the beam, although the force is acting upwards. We note that the deflection at $x = 0.5l$ becomes zero.

The same simulation was performed using the ANSYS finite element simulator to verify the results. The differences amount to less than one percent and are in principle attributable to differences in the numerical solution procedure. The simulations were run on a SUN Sparc 20 workstation. The simulation time for the first case amounted to 91 CPU seconds for Saber and 94 CPU seconds for ANSYS.

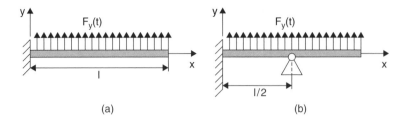

Figure 6.10   Cantilever beam with (a) and without (b) an additional support

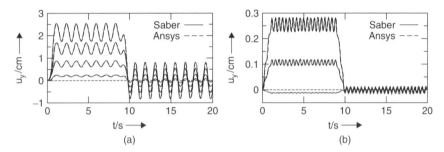

Figure 6.11   Simulation results for the deflection at x = 0.25 l, 0.5 l, 0.75 l and 1.0 l of a cantilever beam with (a) and without (b) an additional support

In the second case with the additional support the times are 155 seconds for Saber and 270 seconds for ANSYS.

Based upon the previous example, it was possible to show that finite elements can be formulated in hardware description languages. The same methodology can also be used for the implementation of other finite elements, such as is also shown in Chapter 8. The calculation using the solver of a circuit simulator does not necessarily demand running times that are higher by orders of magnitude. On the other hand, the approach described above does not form a competition to the regular FE-simulators. The main goal of the work described here remains to bring together electronics and mechanics in order to simplify the design of mixed systems.

6.3.3 Physical modelling

Procedures such as the finite element procedure are certainly the most general solution for the envisaged problem. As a result of the high number of degrees of freedom, problems in the simulation speed occasionally occur. In order to achieve improvements here, for certain geometries — for example, round or square plates — we can give formulae that correspond with a physical modelling. The development of such models requires a considerable degree of modelling effort because it calls for an understanding of the physics of the components.

In what follows, four approaches will be considered in this context. The first possibility is to take a partial differential equation for the mechanical continuum and to represent this using, for example, the method of finite differences on a system of ordinary differential equations, which again can be directly formulated in a hardware description language. The second method relies upon analytical solutions of the partial differential equations in question which are, however, rarely known. Finally, the last two options — the Ritz and Galerkin approaches — attempt to describe bending structures on the basis of a calculus of variations.

Partial differential equations and finite differences

A classical approach to the consideration of the physics of bending structures is to derive a partial differential equation, which can, for example, be represented as a set of ordinary differential equations by the method of finite differences. This step is necessary because analogue hardware description languages cannot in general process partial differential equations directly. The process described was first used by Lee and Wise [224] in order to investigate pressure sensor systems in bulk micromechanics, in which the (quasi-static) solution was built into the respective circuit simulator. In [322], [323] and [324] Pelz *et al.* transferred this solution from the tool level to the model level, where the automatic translation of partial differential equations (in one dimension) into hardware description languages and equivalent Spice net lists was investigated in particular. Consideration was also given to mechanical kinetics. Mrčarica *et al.* [278] also use this approach to consider two-dimensional, partial differential equations, favouring a direct formulation in the in-house hardware description language AleC++. Finally, Klein and Gerlach [195] break up a bending plate into fragments in their approach, and models in an analogue hardware description language are then applied to each of these. These can again be connected to a circuit simulation, thus facilitating the co-simulation of continuum mechanics and electronics. The formulation leads to a system model that is mathematically equivalent to the method of finite differences.

For illustration, the circular plate of a capacitive pressure element will be considered here, see Figure 6.12 and [322], [323] or [324]. A comprehensive description of this example, which will be used frequently in what follows, is found in Section 8.2. The plate is deflected by an external pressure and thus changes the capacitance of the pressure element, which again is detected by a read-out circuit.

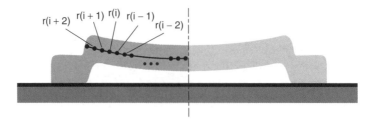

Figure 6.12   Finite differences for a capacitive pressure element

The bending of such a plate can be described by the following partial differential equation, see Gasch and Knothe [113]:

$$\frac{\partial^2 u}{\partial t^2} = -\frac{Et_p^3}{12\rho(1-\nu^2)} \left(\frac{\partial^4 u}{\partial r^4} + \frac{2}{r}\frac{\partial^3 u}{\partial r^3} + \frac{1}{r^2}\frac{\partial^2 u}{\partial r^2} \right) + w \qquad (6.48)$$

Where u is the deflection, E the modulus of elasticity, t_p the thickness of the plate, ρ the density of the plate material, ν Poisson's ratio, w the excitation, and r the (radial) position variable. This is then discretised over the range of the plate radius in n nodes, see Figure 6.12. The above equation is used for each of these n points, whereby the positional derivation is replaced according to the following plan:

$$\frac{\partial u}{\partial r} \approx \frac{1}{12h}(u_{r(i-2)} - 8u_{r(i-1)} + 8u_{r(i+1)} - u_{r(i+2)})$$

$$\frac{\partial^2 u}{\partial r^2} \approx \frac{1}{12h^2}(-u_{r(i-2)} + 16u_{r(i-1)} - 30u_{r(i)} + 16u_{r(i+1)} - u_{r(i+2)})$$

$$\frac{\partial^3 u}{\partial r^3} \approx \frac{1}{2h^3}(-u_{r(i-2)} + 2u_{r(i-1)} - 2u_{r(i+1)} + u_{r(i+2)}) \qquad (6.49)$$

$$\frac{\partial^4 u}{\partial r^4} \approx \frac{1}{h^4}(u_{r(i-2)} + 4u_{r(i-1)} + 6u_{r(i)} - 4u_{r(i+1)} + u_{r(i+2)})$$

As a result of the form of the terms used, the necessity arises to add two further nodes at both ends of the discretised range. These do not describe a real expansion of the plate but can, however, be used in addition to the boundary nodes for the formulation of the boundary conditions. This yields a description with $2n + 4$ degrees of freedom, some of which are dispensed with due to the boundary conditions.

Overall, the drawing up of the equation system and its description in a form compatible with the electronics is definitely specified, but it is very cumbersome to achieve manually. For this reason a model generator is used in [322], [323] or [324], which automatically converts the partial differential equation into a formulation in an analogue hardware description language or a Spice compatible equivalent circuit on the basis of general integrators. The procedure is so general that it can also be used on other partial differential equations such as the heat conduction equation, see for example Bielefeld *et al.* [35] and [36]. However, one remaining limiting factor is the fact that the method is only suitable for relatively simple structures due to the nature of the underlying partial differential equations. Furthermore, in this model only the plate is considered and not its suspension.

Analytical modelling

For some structures, such as square or circular plates, analytical solutions to the partial differential bending equations are known. Models can be created on this basis if the geometric form of a micromechanical structure permits. This is particularly true for very simple structures. Thus Chau and Wise [67] and Bota *et al.* [41], for example, use analytical equations for the modelling of the square membrane of a pressure sensor. In addition to bending mechanics, torsional mechanics can also be considered analytically, as Gómez-Cama *et al.* [122], for example, demonstrate for a capacitive acceleration sensor and Wetsel and Strozewski [428] demonstrate for a micromirror.

To illustrate analytical modelling, the example of a capacitive pressure sensor, see Figure 6.12, will be considered again in what follows. The bending of the upper plate can be described by the following equation, see Timoshenko and Woinowski-Krieger [401] or Voßkämper *et al.* [417]:

$$\Delta\Delta u = \frac{1}{D}(p + p_{el}), \qquad \text{where } D = \frac{E}{1 - v^2}\frac{t_p^3}{12} \qquad (6.50)$$

where Δ represents the Laplace operator, u the vertical deflection, D the bending resistance, p the external pressure and p_{el} an electrostatic pressure caused by the read-out voltage applied through the plates. The bending resistance is again defined as shown via the modulus of elasticity E, Poisson's ratio v and the thickness of the plate t_p. The electrostatic pressure can be described as follows using the radius:

$$p_{el}(r) = \frac{1}{2}\varepsilon_0\varepsilon_{r,eff}(r)\left(\frac{U}{t_c + t_i - u(r)}\right)^2 \qquad (6.51)$$

with the dielectric constants ε_0 and $\varepsilon_{r,eff}$, the radius r, the read-out voltage U, the thickness of the hollow space t_c and the insulator thickness t_i. A direct execution of the four-fold integration of (6.50) for the solution with respect to deflection is not possible because the electrostatic pressure in (6.51) is itself dependent upon the deflection. A polynomial approximation of p_{el} solves the problem, see [417]:

$$p_{el}(r) \approx \sum_{i=0}^{n} a_i r^i \qquad (6.52)$$

The general solution of equation (6.50) is then calculated as:

$$u = \frac{1}{64}\frac{p}{D}r^4 + \frac{1}{4}C_1 r^2\left(\ln\frac{r}{R_0} - 1\right) + \frac{1}{4}C_2 r^2 + C_3 \ln\frac{r}{R_0}$$
$$+ C_4 + \sum_{i=0}^{n}\frac{a_i}{(i+2)^2(i+4)^2}r^{i+4} \qquad (6.53)$$

with the radius of the plate R_0 and four constants C_1 to C_4 that have been yielded by the integration, the values of which are to be determined from the boundary conditions. With the aid of the resulting equation, further effects can be built in, such as the restriction of the plate movement through the insulator, the influence of plate suspension, or the dynamics of the movement.

Ritz method

A further procedure for the modelling of strains is the Ritz method, see for example Bathe [19]. In this process the partial differential equation is solved and an attempt is made to approximate an unknown displacement function, e.g. the deflection of a beam over its length by a linear combination of n interpolation functions. These must each correspond with the geometric boundary conditions. The n coefficients of the interpolation functions are yielded by the requirement that the elastic potential must be minimal. From this, n equations are found, which set the partial derivative of the elastic potential with respect to the coefficients equal to zero. So n equations are available for n coefficients. It should also be noted that the interpolation functions are defined over the entire mechanical structure, which makes the consideration of irregular structures considerably more difficult. The same applies for nonhomogeneous distributions of mass and stiffness. For this reason the significance of the Ritz procedure lies not so much in its direct application, but rather in the fact that it forms the basis of the finite elements method. Nonetheless, the direct use of the Ritz procedure can make sense in some cases.

Galerkin method

As in the finite differences approach, this method, see for example Bathe [19], also generates a set of ordinary differential equations from a partial differential equation:

$$L[\phi] = r \tag{6.54}$$

Where L is a linear differential operator, ϕ the sought-after solution, and r the excitation function. The solution of the problem should correspond with the following boundary conditions B_i:

$$B_i[\phi] = q_i|_{\text{at the boundary of } S_i} \tag{6.55}$$

A prerequisite here is that L is both symmetrical (6.56) and positive definite (6.57).

$$\int_D (L[u])v \cdot dD = \int_D (L[v])u \cdot dD \tag{6.56}$$

$$\int_D (L[u])u \, dD > 0 \tag{6.57}$$

Where u and v are arbitrary functions and D is the range of the operator. The solution should now be approximated as a linear combination of weighted interpolation functions h_i:

$$\overline{\phi} = \sum_{i=1}^{n} a_i h_i \tag{6.58}$$

The h_i interpolation functions are selected such that they each fulfil the boundary conditions. Then the residuum R is calculated as follows:

$$R = r - L \left[\sum_{i=1}^{n} a_i h_i \right] \tag{6.59}$$

For the exact solution the residuum is zero and, for the approximation, should at least be sufficiently low at all points of the solution range. Then the weighting factors a_i can be determined during the approximation of the partial differential equation. For the Galerkin method the following equations are used as the basis:

$$\int_D h_i R \, dD = 0 \qquad i = 1, 2, \ldots, n \tag{6.60}$$

Where D is again the solution range.

Hung *et al.* [156] use the Galerkin method to investigate a pressure sensor, which consists primarily of a bending beam that is fixed at both ends. A voltage and consequently an electrostatic force is applied to this. The time that passes before the beam 'snaps into place' as a result of the positive feedback of the electrostatic force, i.e. forcefully rests upon the insulator, is strongly dependent upon the prevailing air pressure. The modelling uses the Euler equation for bending beams and Reynolds' equation for air damping. The authors use the Galerkin method with up to four interpolation functions, which are determined with the aid of a FE simulation. They thereby achieve an acceleration of the simulation by a factor of between 4 and 105 in comparison to FE simulators, with deviations from the FE simulation in the range of 1%–14%.

This method permits the formulation of lower-order models. However, it requires that the system can be considered as a comparatively simple structure, because the starting point, the partial differential equations and boundary equations, either cannot be set or can be set only with great difficulty.

6.3.4 Experimental modelling

Introduction

Experimental modelling dedicates itself to the creation of models on the basis of measured data or FE simulations. The internal physics of the components is disregarded and only the terminal behaviour considered. In this manner we obtain so-called macromodels that can be simply formulated in a hardware description language. We thus obtain efficient and numerically unproblematic models. This method has its advantages if it is difficult or even impossible to derive the physical background of a component. However, its main problem is that the resulting models are only valid for precisely one geometric form of the structure and set of technology parameters. Every change means that a new model must be drawn up.

A whole range of approaches extract the main corner-stones of the behaviour of a component from measurements or simulations using finite elements and use this for simple models consisting of few equations, see for example Ansel *et al.* [11], Hofmann *et al.* [149], [150], Karam *et al.* [179] and Nagel *et al.* [292]. In what follows three approaches will be considered that aim in the aforementioned direction.

Table models

The simplest case of experimental modelling is based upon a list of input and output values, thus arriving at a table model that only considers the static case. In this manner it is possible, for example, to draw up a table listing pressures and the associated capacitance values for the pressure elements described above. Such table models lead to characteristics with kinks that can considerably detract from the convergence of the simulator. This problem can be circumvented by using the present value pair as a support point for the characteristic, e.g. on the basis of splines, which typically removes the numerical problems. In this manner measured values can be very simply integrated into a simulation. More elaborate procedures estimate the structure of the equations and move themselves to the identification of the associated parameter.

Identification of a harmonic oscillator

In [11], Ansel *et al.* consider a seismic acceleration sensor as a harmonic oscillator. For the modelling a linear differential equation is used for the force f and the deflection x:

$$a_0 f + a_1 \frac{df}{dt} + \cdots + a_m \frac{d^m f}{dt^m} = b_0 x + b_1 \frac{dx}{dt} + \cdots + b_n \frac{d^n x}{dt^n} \qquad (6.61)$$

For a spring-mass system, for example, m is set to 0 and n to 2. Here b_0 represents the spring constant, b_1 the viscous damping, and b_2 the seismic mass. For the system currently under consideration the parameters a_i and b_i are automatically obtained from the results of a simulation using finite elements. For this purpose the classical methods for system identification are used. This describes the mechanical section of the system. In addition, there is the conversion of mechanical deflection into capacitance based upon an interlacing comb structure. A table model is used for this, which is also determined on the basis of simulations using finite elements.

General identification

Hofmann *et al.* [149] and [150] propose a general procedure in order to put together the behaviour of a component from functional modules. The modelling is based upon a FE model, the behaviour of which is stored in a macromodel. Thus the complexity and nature of the underlying (partial) differential equations are not known in advance, so that we have to start from the assumption of the existence of strong couplings and nonlinearities. Furthermore, it is required that inputs and outputs of the FE model can be formulated in an integral manner, i.e. they are not position dependent.

We now start with a basic model, the parameters of which should be identified with the aid of various optimisation procedures. For oscillating systems, for example, equation (6.61) would be a good starting point, whereby the parameters a_1 and b_1 would have to be determined. For the general case these can be determined from the criterion that the resulting model should behave as closely as possible to the FE model. The target function of optimisation is thus the minimisation of the behaviour difference between the predetermined and sought-after model, i.e. [150]:

$$\sum_j^{\#outputs} \sum_i^{\#timesteps} (f_{FEM_i}(t_j) - f_{MACRO_i}(t_j))^2 \tag{6.62}$$

For optimisation, gradient procedures, simulated annealing, or genetic algorithms can be used and it is also possible to switch between these. The resulting parameters initially apply only for the selected input function. In [150] it is thus proposed to initially define a set of input functions, which represent reality as well as possible. Then optimisation takes place primarily for the input function, the macro model of which exhibits the greatest differences in relation to the FE model. This procedure corresponds with a parameter identification for nonlinear systems.

Now, if it is difficult to arrive at an acceptable solution using parameter optimisation, this raises the question of whether the assumptions with regard to the structure of the solution equations were correct. Once again, the problem lies in the nonlinearities that rule out an analytical solution of the problem. The solution proposed by Hofmann *et al.* consists of setting operators that evaluate the differences between the FE and macromodel, such as for example 'rate of rise too low', 'overshoot too low' and so on. On the basis of this information a fuzzy controller base decides on possible structural changes. So we now go from parameter identification to system identification.

Overall, the procedure supplies efficient and numerically unproblematic models, that can be easily formulated in hardware description languages, e.g. HDL-A [149]. However, a significant computing time has to be expended for model generation. Furthermore, the validation of the generated models remains difficult, since firstly the quality and coverage of the selected input functions is sometimes questionable and secondly the inner physical structure is not available for an investigation into the plausibility. Finally, this type of modelling has to be performed afresh for virtually every variation of the micromechanical geometry or the underlying technology.

6.4 Summary

In this chapter, methods for the modelling of multibody mechanics and continuum mechanics have been highlighted and the representation of the resulting models shown in hardware description languages. This, along with the results of the previous chapters, facilitates a full, universal modelling of mechatronic and micromechatronic systems in hardware description languages.

Now that the basic technologies have been dealt with in the preceding chapters, the following two chapters on mechatronics and micromechatronics supply a range of demonstrators to illustrate their application.

7

Mechatronics

7.1 Modelling of Mechatronic Systems

The aim of this chapter is to apply the foundations obtained in previous chapters to actual mechatronic systems. The interaction between the domains is particularly significant here because the interfaces contribute significantly to system behaviour. In particular, we are aiming too low if we only consider electronics or mechanics independently of each other. The problem of the joint simulation of electronics and mechanics must be solved, which again throws up a whole range of problems:

In the case of mechatronics, the time constants of mechanics and electronics often differ by orders of magnitude. For macromechanics we can expect oscillations of a few (tens of) hertz. In electronics the figure lies four to six orders of magnitude higher. So we could assume that the dynamic interaction between electronics and mechanics can be disregarded. The opposite is true. For example, a wide range of control algorithms are performed on embedded controllers. Their running time again lies in the millisecond range, so that dynamic feedback between electronics and mechanics very definitely plays a role. This requires the dynamic simulation of the entire system in order to be able to track cyclical dependencies, including those that cross domain boundaries. Another reason for the importance of this is the fact that domain boundaries often also represent the interfaces between design teams working in parallel.

For the field of mechanics, precise models that are compatible with an electronics simulator must be prepared. During the following chapter we will exclusively consider multibody mechanics, which is generally sufficient for system considerations in the field of macro mechatronics. Even with this limitation the vectorial nature of mechanics is not easy to represent on a circuit simulator.

An efficient conversion is of crucial importance for the field of software in particular. Millions of machine instructions are performed in a single second of real time. On the other hand, it is necessary to precisely determine the timing of the functions implemented using software, which requires a precise synchronisation between software and electronics. This is indispensable in order to correctly reflect the dynamics between software, electronics and mechanics.

Mechatronic Systems   Georg Pelz
© 2003 John Wiley & Sons, Ltd   ISBN: 0-470-84979-7

In addition to the provision of the models and the dynamic simulation of a system that goes beyond domain limits, the representation of the results can sometimes be a problem. Of course, we always obtain the values of system variables plotted against time, as is also normal for electronics simulation. In the case of mechanics, however, we would often prefer an animation, in order to be able to evaluate the system behaviour at a glance. As far as software is concerned, the typical outputs of an electronics simulator are virtually useless. We would like a debugger, like those used in pure software development, which illustrates the sequence of the software and furthermore permits control of the sequence, perhaps by breakpoints.

As will be shown in what follows, the introduction of hardware description languages into mechatronics lags behind that of microsystem technologies. A significant reason for this is that microsystem technologies developed from microelectronics, so hardware description languages, which were initially developed for microelectronics, were quickly implemented there too. By contrast, mechatronics developed from mechanical engineering, where electronics is often reduced to control technology and can thus be considered using comparatively simple equations. Recently, hardware description languages have also begun to be encountered in mechatronics, with the automotive industry taking the lead.

The use of hardware description languages for the design of mechatronic systems will be illustrated on the basis of four examples. These are a semi-active wheel suspension system, an internal combustion engine with drive train, a camera winder and a disk drive.

7.2 Demonstrator 1: Semi-Active Wheel Suspension

7.2.1 System description

In what follows a semi-active wheel suspension will be described, see Hennecke *et al.* [137] or [138], Duttlinger and Filsinger [89] or Roppenecker [352] for the technical fundamentals. The idea of semi-active wheel suspension is that the system adapts the parameters of shock absorbers or body springs to the current road conditions and the corresponding driving situation. This is achieved by an embedded processor, which means that electronics, mechanics and software have to be considered at the same time here. The system described in what follows reflects the concept of BMW's 'electronic damper control', see [137] and [138] and Figure 7.1. A similar system is also offered by Mercedes-Benz, see [89]. The difference between semi-active and active wheel suspension is that, in the latter case, forces can be applied by hydraulics, for example, in order to improve driving safety and comfort. Put simply, the vehicle lifts one or more wheels up in order to minimise the vertical movement of the body. This approach has, however, not yet become prevalent for reasons of cost and energy.

The motivation for the use of semi-active wheel suspension is that driving safety and comfort represent competing goals in the context of suspension systems. Driving safety is jeopardised because unevenness in the road triggers vibrations in the wheel or body. The former lie in the range of around ten hertz, the latter in the range of one hertz. In both cases it should be ensured that these resonances are sufficiently damped. Otherwise, strong variations in the wheel load will arise. In particular, the wheel load will fall significantly as a result of the upward movement in question. In the extreme case individual wheels will lift. The frictional connection between the wheels and the road, and consequently the cornering force of the wheels in question, then falls to zero, which can have severe consequences in a bend. These problems could be countered by basically setting the damping level high. However, this would have a negative effect upon the ride comfort, which would transmit every unevenness in the road directly to the body and thus to the passengers. This is particularly true for the vibrations in the range of four–eight hertz, which would be perceived as unpleasant or at least uncomfortable by the passengers.

Whilst an electronic control of the wheel suspension was not possible the primary issue was to find a reasonable compromise between safety and comfort. Semi-active wheel suspension is based upon the principle that the damping of road conditions can be switched over appropriately during the journey. Problematic driving situations requiring a high level of damping are recognised firstly by the consideration of the vertical acceleration of the body, from which the triggering of vibrations by the road can be deduced. In addition, driving manoeuvres that also require increased stability are recognised, for example, sharp braking, fast driving through bends or quick accelerator pedal movements in automatics. The resulting pitching and rolling movements of the vehicle should be limited by higher damping. After the identification of the road condition and the driving state, the next step is to determine the correct damping level and to set this at the shock absorber.

The implementation is based upon a digital controller that processes embedded software, see Figure 7.1. This carries out the actual control algorithm and takes on a whole range of additional functions such as, for example, the plausibility testing of the sensor values to be processed, the safety concept, or the provision of data to other components of the vehicle. In addition, there are electronics for the signal processing, such as D/A and A/D converters, which provide the connection between the digital and analogue worlds. The actual conversion between the physical quantities is taken care of by the acceleration sensor and the adjustable shock absorbers. Finally, the mechanics of the suspension also has to be taken into account.

The central component of the system is the regulateable shock absorber, which will be considered in more detail in what follows. Fundamentally, shock absorbers use the Stokes' friction of a viscous liquid, e.g. hydraulic oil, in order to convert motion energy into heat. The oil is squeezed through a narrow valve in accordance with the movement. If we control the route of the oil in accordance with the direction of flow, then different valve diameters and thus different damper constants can be provided for compression and tension mode. Furthermore, several damper

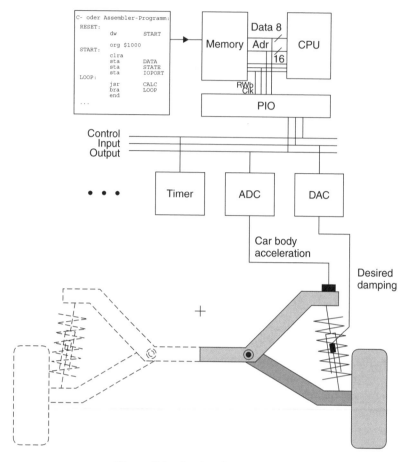

Figure 7.1   Semi-active suspension

characteristic lines can be planned for the adjustable shock absorber. By using a
rotary disk valve the route of the hydraulic oil is determined such that the oil passes
through various valves. The power consumption is thus limited because only the
rotary disk valve has to be operated. Typical delay times for the adjustment of the
characteristic lie in the range of around 20 ms.

7.2.2   Modelling of software

The software in this demonstrator runs on an embedded processor, which is compatible with the Motorola 68HC05. It was developed in C and then initially compiled
into Assembler and from there compiled into machine code. The basis of the mechatronic HW/SW co-simulation used here is software modelling, which has already
been described in detail in Chapter 5 and [326], [328]. For the sake of simplicity,
only the dependency between the body acceleration and the damper characteristic

will be implemented here. More complex algorithms and further input data can be added in a straightforward manner.

7.2.3 Modelling of mechanics

The approach to the modelling of the mechanical components depends to a large degree upon the desired application of the models. Thus, we can provide complex multibody models for the development of the mechanics which can be used, for example, to determine the tilt of the wheels or changes to their toe-in in relation to the spring deflection of the car body, see for example Schmidt and Wolz [365]. However, these details are of less importance to the development of the electronics. In this context the overall behaviour of the system is much more interesting, and simpler models are sufficient for the consideration of this.

The model of the wheel suspension is based upon the fundamental model presented in Chapter 6. However, in this context some further boundary conditions have to be taken into account. Firstly, we should note the fact that real shock absorbers exhibit different characteristics for the compression and tension modes. Typically the damping is significantly higher in tension mode. This is because the dampers should transmit road unevenness to the passengers as little as possible. This means that comparatively low forces should arise in the compression mode. Consequently most of the consumption of the motion energy occurs in the tension mode. This situation generates a first nonlinearity. A second naturally arises as a result of the switching of the damper characteristics, which is precisely the main purpose of the system.

It has already been demonstrated in Chapter 6 how such a model is put together from the basic models for masses, springs, dampers, etc. for a quarter of a car. For this reason, a system-oriented approach to modelling will be followed here. The Lagrange equations shall form the basis for this. As shown in Chapter 6 these take the following form:

$$\frac{d}{dt}\left(\frac{\partial T}{\partial \dot{q}_i}\right) - \frac{\partial T}{\partial q_i} = Q_i \qquad (i = 1, 2, \ldots, n) \tag{7.1}$$

The degrees of freedom are again the y-positions. The kinetic energy of the system T is found from the kinetic energy of the two masses:

$$q_1 = y_a, q_2 = y_b, \qquad T = \tfrac{1}{2}m_a\dot{y}_a^2 + \tfrac{1}{2}m_b\dot{y}_b^2 \tag{7.2}$$

The generalised forces are the spring, damper and weight forces:

$$Q_{a1} = k_w(y_s - y_a + l_{0a}),$$
$$Q_{a2} = -k_s(y_a - y_b + l_{0b}),$$
$$Q_{a3} = -b(\dot{y}_a - \dot{y}_b),$$

$$Q_{a4} = -m_a g,$$

$$Q_{b1} = k_s (y_a - y_b + l_{0b}),$$

$$Q_{b2} = b(\dot{y}_a - \dot{y}_b),$$

$$Q_{b3} = -m_b g \tag{7.3}$$

where k_w and k_s denote the constants of wheel and body springs, Q_{ai} and Q_{bi} the components of the generalised forces Q_i on the bodies A and B, and b the coefficient of damping. Furthermore, the convention is used for the springs that a positive force increases the positions in question. The constants l_{0a} and l_{0b} correspond with the y-positions of the two related bodies in a relaxed state. At the damper, positive forces bring about an increase in the coordinate difference. Thus the damping force tends to resist a positive relative velocity. The weights Q_{a4} and Q_{b3} finally effect a reduction in the positions and are thus counted negatively. Substitution into the Lagrange formula gives the following equation system:

$$m_a \ddot{y}_a = k_r (y_s - y_a + l_{0r}) - k_f (y_a - y_b + l_{0f}) - b(\dot{y}_a - \dot{y}_b) - m_a g$$

$$m_b \ddot{y}_b = k_f (y_a - y_b + l_{0f}) + b(\dot{y}_a - \dot{y}_b) - m_b g \tag{7.4}$$

This can be directly formulated in an analogue hardware description language, whereby the damper constant b can be set to the value in question using an if-then-else construct.

7.2.4 Simulation

The test case for which the simulation should be performed is, as in Chapter 6, driving over a step of 5 cm height. As described in Chapter 6, the model of a quarter of a car can be used again at this point. The first result of the simulation is shown in Figure 7.2 and represents the relative movement of wheel and vehicle body after driving over the bump. Initially the wheel starts to move and compresses the body spring. This movement produces an overshoot. Overall, we recognise that the natural frequency of the wheel actually lies in the range of around ten hertz. The oscillation, however, decays very rapidly as a result of the damping, which is increased at the change-over point. The car body follows the vertical movement at a significantly slower rate than the wheel. The forces in question are applied by the compressed body spring. Here too the natural oscillation is, as expected, set to a value of around one hertz. The change-over of the damping from the 'soft' to the 'hard' characteristic is triggered by the digital signal shown at the top and takes place around 50 ms after the step is reached.

The difference in driving behaviour resulting from the change-over is particularly well illustrated by the consideration of the body spring compression, which is equated with the compression of the shock absorber. This is shown in Figure 7.3 for the same case of driving over a step of 5 cm height. In the first 50 ms there

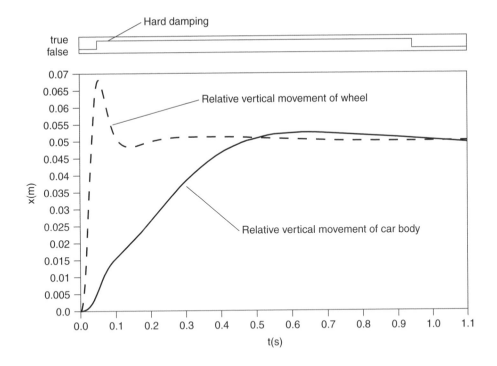

Figure 7.2 Relative vertical movement of wheel and body

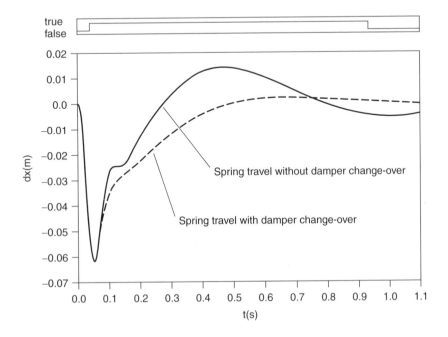

Figure 7.3 Spring compression with and without the changeover of the damper characteristic

is still no difference, because the change-over to hard damping has not yet taken place. Then we see clear differences. If the change-over does not take place, then the spring compression goes positive, which means that the car rises off the springs and thus the wheel loading falls. This leads to the safety problems discussed above. When the damper characteristic is changed over, the spring compression exhibits a significantly more desirable behaviour.

As discussed above, it is also necessary to concern ourselves with the visualisation of the mechanics and the software. For the mechanics this quickly becomes an ad hoc solution, unless CAD data is available from the development of the mechanics, which can be drawn upon for a suitable representation. The solution shown here shows a graphic representation of a quarter of a car, which is 'driven' by the simulation data, see Figure 7.4. This functions both as direct output during simulation and also for a subsequent consideration of the extracted data.

By contrast, a general solution can be found for the software. As already indicated it is a question of considering the software using a type of debugger and controlling its processing, see Figure 5.7. In contrast to the tools normally used, this is a debugger that considers the processing of software on virtual hardware. Naturally, if this is to run properly an exact synchronisation between hardware and software is crucial. This is guaranteed by the approach to mechatronic hardware/software cosimulation described in Chapter 5. This synchronisation ensures that when a break-point is reached, the simulation of mechanics and electronics

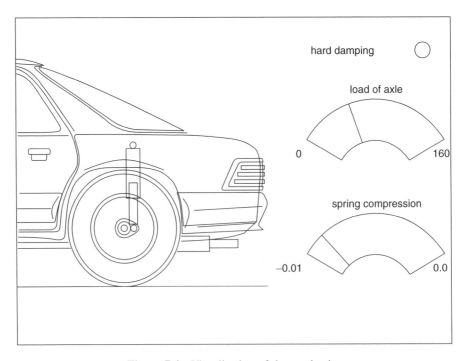

Figure 7.4   Visualisation of the mechanics

do not proceed further. Such a debugger has been implemented and used. It can represent the underlying software both on assembler and high-level language level. The user interface of the debugger is shown in Figure 5.7.

The models were formulated in the analogue hardware description language MAST and simulated using the Saber electronics simulator. On a SUN Sparc 20 workstation the simulation requires 22 CPU seconds. This includes the simulated processing of 3.8 million machine instructions of the embedded processor.

7.3 Demonstrator 2: Internal Combustion Engine with Drive Train

7.3.1 System description

This example relates to the modelling of the propulsion of a motor vehicle see also [332]. If all components that are relevant in this context are to be considered, then the modelling would extend from the accelerator pedal, via the actual engine and the drive train, to the road, which has a certain gradient. If we impose narrower limits on the model, then we can at best investigate specific parts of the system. In what follows, the overall system will be investigated, whereby the necessary foundations can be found for example in Roduner and Geering [349], Hockel [146], Tiller *et al.* [400] or in the publication by Bosch [39], [40]. As in the previous example, generic models are used where possible, which can be adapted to the application case in question by suitable parameterisation. Again, the model is based upon the assembly of the overall model from basic components.

The system under investigation is shown in Figure 7.5. The partitioning of the system for the modelling mainly follows the function blocks shown in the illustration. At the top level, the system model consists of an electronic circuit diagram. The components of this have been modelled using an analogue hardware description language, and so the whole of the overall system can be simulated without further problems. Particular mention should be made of the fact that two types of modelling have been used here.

The drive train primarily relates to the rotational and translational movements, whereby here a line represents the duality of torque / angular velocity or force / velocity. Kirchhoff's laws in particular apply here, i.e. the forces and torques at a node and the velocities and angular velocities in a closed loop add up to zero. The lines in question are printed in bold.

All other lines are directional. They form a network that is comparable to the block circuit diagrams of control technology. In particular, Kirchhoff's laws do not apply here. The main task of this partial network is to supply the torque generation block and the automatic gearbox with the necessary data. This includes the control of the combustion engine.

In the system description we start with the accelerator pedal, which in our case represents the driver's primary input. Its angle a_{pe} sets the kickdown mode in train

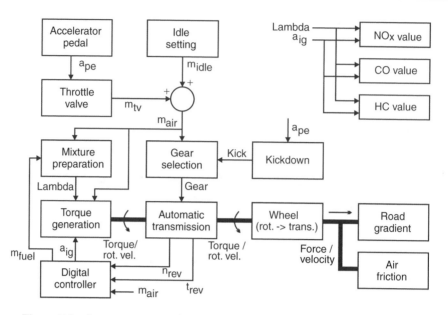

Figure 7.5   System structure of the internal combustion engine with drive train

when the driver stamps upon the accelerator pedal, which may cause the automatic gearbox to change down so that a higher acceleration can be achieved. Furthermore, a_{pe} determines the position of the throttle valve, which again provides a suitable quantity of air m_{tv} for combustion. In addition, there is a further proportion of air m_{idle} which corresponds with the idle setting and is permanently available. The sum of the two air masses m_{air} is firstly a measure for the desired acceleration and is therefore fed into the gear selection and the digital computer. Secondly, m_{air} is naturally available to the mix preparation, which calculates the fuel mixture ratio λ from the air and fuel mass m_{fuel}. The digital micro-controller determines the fuel quantity and also the ignition advance angle a_{ig}. Its inputs are the engine speed n_{cs}, a trigger signal of the crankshaft t_{cs} and the air mass m_{air}. The regulator can be described as a software model, see Chapter 5, or as a simple behavioural model.

The actual engine is represented by the torque generation block, which converts the fuel mixture ratio λ, the air mass m_{air} and the ignition advance angle a_{ig} into the mechanical duality of torque and angular velocity. The first step in the modelling process is to estimate the energy of the fuel mass and to multiply this by the associated efficiency. Furthermore, this raw torque is further reduced if the engine is not run in the optimal range of λ, m_{air} and a_{ig}. Then the rotary motion is transmitted to the gearbox and there suitably converted, which leads to a different defined relationship of torque and angular velocity. The model of the wheel converts the rotary motion into a translational motion, which is characterised by the duality of force and velocity. Counter-forces come into play here, which represent the gradient of the road and the drag due to air resistance. Finally, the exhaust blocks supply a broad overview of the current NO_x, CO and HC values of the engine, which are obtained from λ and a_{ig}.

7.3.2 Modelling

At this point some of the above-mentioned blocks will be described. We begin with the generation of torque by an internal combustion engine. The underlying process represents the conversion of chemical into mechanical energy. The input quantities are the air mass m_{air}, the fuel mixture ratio λ, the ignition advance angle a_{ig}, and the engine speed at the crankshaft n_{cs}. We could start by deriving the resulting torque M from the average pressure in the cylinder p_m and the piston-swept volume V_D, see [39]:

$$M = \frac{V_D \cdot p_m}{2\pi} \tag{7.5}$$

However, the average pressure in the cylinder cannot easily be determined from the input quantities. Rather, p_m depends upon the combustion process, which takes place in three spatial dimensions. Also problematic is the irregular shape of the combustion space, which furthermore continuously changes shape. Finally, turbulence effects in the mixture of air and fuel often cannot be disregarded. Thus the physical modelling is separated from the consideration of the system as a whole. Since in this case there is no good foundation for a structural modelling, we now turn our attention to experimental modelling. Figure 7.6 shows a typical characteristic of torque against engine speed for a predetermined air mass, see [39]. Such a characteristic can be described by an inverted parabola, whereby its parameters are to be adjusted for each operating case of the engine under consideration. In this context a range of values for different air masses should be recorded.

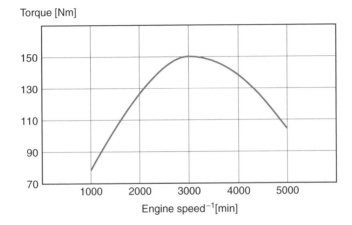

Figure 7.6 Typical characteristic of torque against engine speed for a given air mass

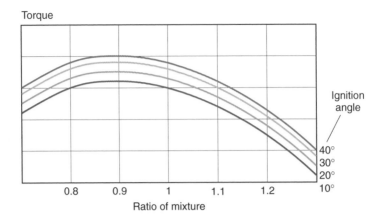

Figure 7.7 Typical dependency of the torque upon the fuel mixture ratio and the ignition advance angle

In this manner a 'raw torque' figure is obtained that can be adjusted by multiplying it by a weighting factor to take account of the other input quantities λ and a_{ig}. We thus do not have to store the characteristics for all sensible combinations of m_{air}, n_{cs}, λ, and a_{ig}. The factor for the calculation of the actual torque is found from Figure 7.7, which shows a typical dependency of the torque upon the fuel mixture ratio and the ignition advance angle, see [39].

The generated torque is opposed by a corresponding counter-moment, which is found from the inertial force of the vehicle mass and the forces due to air friction and road gradient[1]. The rolling friction is disregarded in this context. The inertial force F_i can be replaced by:

$$F_i = -m\dot{v} \tag{7.6}$$

where m represents the vehicle mass and v the vehicle velocity. For the drag F_l:

$$F_l = -\tfrac{1}{2}c_W A\rho v^2 \tag{7.7}$$

where c_W is the drag coefficient of the car body, A its frontal area and ρ the density of the air. The force from upward and downward gradients, F_g, is formulated:

$$F_g = -mg\sin(\alpha) \tag{7.8}$$

where g is gravity and α the angle of the road gradient. A single hardware description is used for each of these three force components. This is connected such that the counter-forces are summed. The wheel model converts the counter-force into a counter-moment which again is linked to the generated moment via the gearbox.

---

[1] Rise or fall.

7.3.3 Simulation

The test case that underlies the simulation is a virtual ride of 35 seconds real time, see Figure 7.8. From top to bottom the following values are represented: the operation of the accelerator pedal as a percentage of the maximum angle; the automatically selected gear; the engine speed and the road speed. Regarding the road profile, the vehicle comes upon a gradient of 10% after 10 seconds, which returns to a level section after 25 seconds.

At the beginning of the simulation, the driver operates the accelerator pedal at 20%, so that the vehicle slowly accelerates away from stationary. Upward gear changes are made until the vehicle is in third gear. The engine speed lies at just under 2000 rpm. The road speed rises to around 50 km/h. After ten seconds the vehicle reaches the incline, which after a short time results in a slight decrease in road speed. The driver reacts to this by pushing the accelerator pedal down to 90%. The gearbox, which in the meantime has changed up into fourth gear, reacts and changes through third gear into second gear. The engine speed rises accordingly and the vehicle accelerates up the hill. With the higher speed the gearbox can change back up into third gear after around 18 seconds, which again reduces the engine speed slightly. After 21 seconds the driver lifts off the accelerator a little, so that the speed falls from approximately 80 km/h to 70 km/h. At 25 seconds the incline is ended, so that the vehicle easily accelerates at 60% throttle in fourth gear and at 35 seconds a road speed of around 90 km/h is achieved.

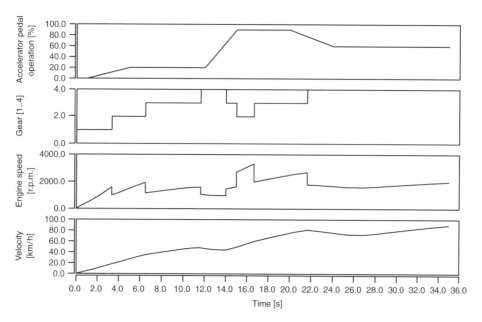

Figure 7.8 A virtual ride by car. The curves represent from top to bottom: accelerator pedal as a percentage of full acceleration; automatically selected gear; engine speed and road speed

7.4 Demonstrator 3: Camera Winder

7.4.1 Introduction

The third example originates not from automotive engineering. It relates to a pho-
tographic camera with an external motor winder, which looks after the automatic
film movement. This system differs from the two other mechatronic examples
because, in this case, the electronics provides the drive and not only the control of
the system. Furthermore, at the time of the investigation not all components of the
system were available. Nevertheless, the system could still be investigated on this
basis, and this investigation yielded a significant foundation for the design deci-
sions under consideration. Thus the described approach is capable of supporting
the top-down design of mechatronic systems. The subject is also dealt with in Pelz
et al. [331] and Landt [213].

The model was set up for this system to illustrate a real system, the Leica R8,
together with its motor winder. At the time of modelling the winder was still at
the development stage. The main task was to maximise the number of pictures
that could be taken by a battery or rechargeable battery. Several motors were
investigated to this end; one of the candidates was not even in existence at the
time. All that was available for this motor was a set of parameters that had been
calculated and estimated in advance. One of the main questions was whether it
would be worthwhile to actually have this virtual motor developed. The gearbox
had to be set up for each motor in order to minimise the power consumption whilst
achieving the specified time for taking a picture.

7.4.2 System description

The system includes camera and motor winder and has an electronic and a mechanical circuit. The electronics consists of a battery pack, the power transistors, the
associated control circuit and the electric motor, which is naturally also part of the
mechanics. This subsystem also comprises of a gearbox, a mechanical load and a
mechanical stop. The mechanical load represents the effect of friction when the
film is pulled out of its cartridge.

Figure 7.9 shows the (simplified) system structure, placing emphasis on the
mechanics and the interface between electronics and mechanics. Again, the simulation is performed using an electronics simulator, which we can expect to correctly
process the electronics.

7.4.3 Modelling

Batteries and rechargeable batteries

The model of the battery or rechargeable battery pack is based upon measurements
of various brands and types. First a load cycle is defined that imitates the typical

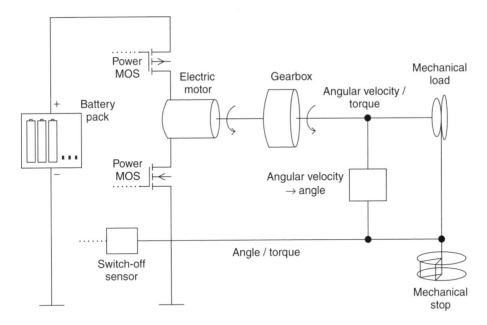

Figure 7.9   System structure of the camera winder

use of the camera. Then a large number of cycles are performed under defined temperature conditions, e.g. $+20°C$ and $-20°C$. The voltage and internal resistance are measured regularly and related to the charge consumed. This information parameterises the model of the battery/rechargeable battery pack, which consists primarily of a voltage source with internal resistance. This can be used to evaluate the variations in the number of pictures that are produced by the temperature and the brand and type of the batteries and rechargeable batteries.

Electric motor

Only direct current motors with permanent magnets will be considered in this context. The modelling, see Hardware description 7.1, is initially based upon equating the sum of the internal voltages with the externally applied voltage. This corresponds with a series connection of armature resistance, armature inductance and generator voltage. This yields the following voltage balance:

$$v_{in} = R_a i_a + L_a i_a + b\omega \tag{7.9}$$

where v_{in} is the input voltage, R_a the resistance of the armature winding, L_a the inductance of the armature winding, i_a the armature current, b the generator voltage constant (back-emf) and ω the angular velocity of the shaft. The resulting torque is found from the generated torque minus the moment of inertia of the armature and the frictional moment.

```
template my_dcpm A1 A2 WRM WRME= kt, b, la, ra, j, dft
 electrical    A1, A2
 rotational_vel WRM, WRME
 number          kt  = 1.0,               # Torque constant
                 b   = 1.0,             # Generator voltage constant
                 la  = 1e-6,             # Armature inductance
                 ra  = 1.0,             # Armature resistance
                 j   = 1.0,              # Moment of inertia
                 dft = 0.0         # Frictional losses, dynamic
{
 branch iin = i(A1->A2),               # Input current, motor
        vin = v(A1,A2)            # Terminal voltage, motor
 branch at = tq_Nm(WRM->WRME),          # Torque at shaft
        av = w_radps(WRM,WRME)     # Angular velocity at the
                                          shaft
 val   tq_Nm frict                  # Frictional moment
 var   nu    av_t, iin_t       # Derivative of av and iin
                                   # with respect to time

     values {
      if (av > 0) {          # Frictional moment acts against the
       frict = dft;                   # rotation direction
      }
      else {
       frict = -dft;
      }
     }
 equations {
   iin_t = d_by_dt(iin)         # Calculation of the derivative
                                       # with respect to time

   av_t = d_by_dt(av)          # Calculation of the derivative
                                       # with respect to time

   vin =
     ra*iin + la*iin_t + b*av               # Armature circuit
     at = kt*iin - j*av_t - frict # Calculation of output torque
 }
}
```

Hardware description 7.1   MAST model of the direct current motor

Gearbox

Like the electric motor, the gearbox represents a basic model that can be used very diversely. Its function is to set the relationship between rotational velocity and torque between the two mechanical terminals according to the transmission ratio. The following equations apply:

$$\omega_A = \alpha \omega_B$$

$$M_B = \eta \alpha M_A \qquad (7.10)$$

where the angular velocities ω_A and ω_B are in a fixed ratio of α, the moments M_A and M_B are in a fixed ratio of $1/\alpha$ and furthermore the efficiency η is taken into account. Thus there is nothing further standing in the way of a direct implementation of the model in VHDL-AMS or another hardware description language.

Mechanical load

A further problem in the modelling is the mechanical load, which arises from the film being pulled out of the cartridge. Here too — as in the case of batteries and rechargeable batteries — it makes little sense to model the underlying physical relationships. Firstly, these processes are complex, such as for example the transition from static to sliding friction. Secondly, the necessary material properties cannot easily be determined. In addition, a physical model would involve significant calculation time, so that it would barely be suitable to investigate the system as a whole. Here too experimental modelling is used, with measurements of the load moment being incorporated into a model. Such a table model breaks down the winding into 18 increments each of $10°$, for example. A complete winding movement thus covers $180°$. Every section is assigned a load moment that opposes the motor moment. This yields an incremental or at best a piece-wise linear path of the load moment characteristic, see Figure 7.10. The characteristic thus has irregularities or at least kinks. Both make the numerical calculation of the underlying differential equation system considerably more difficult. To remedy this, the characteristic was approximated in the form of a Chebyshev polynomial, although spline or Bézier interpolations would also have been feasible.

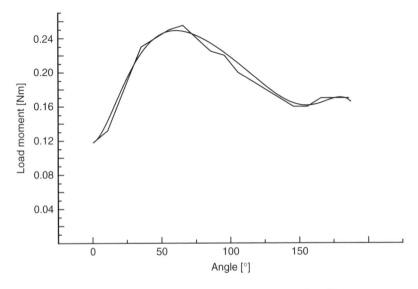

Figure 7.10 Measured and approximated load torque during film movement

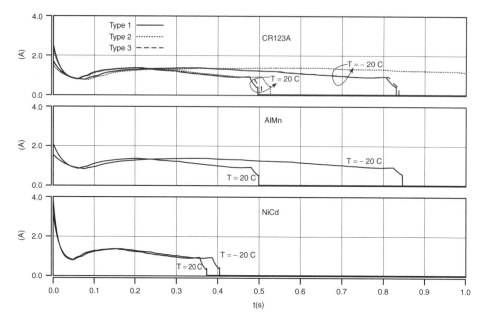

Figure 7.11 Armature currents for various battery/rechargable battery types and temperatures

7.4.4 Simulation

For the simulation shown in the following, see Figure 7.11, a motor was selected and operated using extremely varied batteries and accumulators at $+20°C$ and $-20°C$. All the curves show the armature current, the integral of which is a measure for the charge consumption. When the armature current is zero, the winding operation is complete. Thus we can also use these diagrams to check adherence to the time specification for the winding operation. Individually, lithium batteries (CR123A), AlMn batteries and NiCd rechargeable batteries were used. It is striking that the NiCd accumulators are considerably less problematic, which is due to their overall lower internal resistance. Furthermore, we see that in some cases the lithium batteries are no longer capable of adhering to the specification of 850 ms for a winding operation at low temperatures.

7.5 Demonstrator 4: Disk Drive

7.5.1 Introduction

The section gives an overview of electronics and firmware development using The Virtual Disk Drive, which is a generic model of a disk drive, formulated in hardware description language, see also [78], [320], [321]. The Virtual Disk Drive covers digital, analogue and power electronics, firmware and mechanics. Thus it even allows us to assess functionality spread over all these domains. One example of this is track following and track seeking for the read/write-head.

Moreover, this section details the resulting electronics design methodology. For instance, it will be shown how key system properties, e.g. seek time, can be determined by means of the mixed simulation of mechanics, electronics and firmware. In addition, the same simulation environment is used to realistically verify analogue and digital circuitry as well as firmware.

7.5.2 The disk drive

The following section details some typical configurations of a disk drive and discusses how these translate into requirements for the associated electronics, see also Figure 7.12. A drive normally contains up to five rotating disks. This disk (stack of disks) is driven by the spindle motor, which is a brushless DC motor. The rotational velocity varies between 4200 and 15 000 revolutions per minute.[2]

The RW-head flies on an air cushion 10–50 nm above the disk surface. It is supported by the load-beam, which can be moved about its pivot by the so-called voice coil motor. This consists of a coil that lies in a fixed magnetic field provided by permanent magnets, see Figure 7.13. Any current through the coil results in a torque on the load-beam and thus a circular motion of the RW-head.

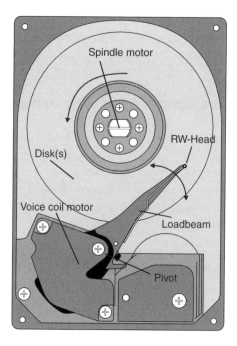

Figure 7.12 Hard disk drive overview

---

[2] Higher as well as lower rotational velocities are chosen at times, e.g. to cope with low latency or low noise requirements.

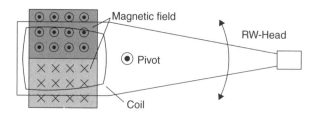

Figure 7.13   Voice coil motor

The data on the disk is organized in circular tracks. The outermost track may be located at a radius of 1.8 inch for a typical 3.5 inch form factor drive.[3] This means that the maximum length of a track is 11.3 inches, which gives rise to a linear speed of about 1880 inch/s for a 10 000 rev/min drive. This, together with a bit density within a track of 400 Kbit/inch, for example, in turn leads to a peak data rate of 750 Mbit/s. Interestingly enough, 1880 inch/s — in terms of the velocity of a car — is far beyond any speed limit, while bits of the length of about 60 nm are reliably read.

The track density of drives currently under design is in the range of 30 000–100 000 tracks per inch. Together with bit densities of 300–500 Kbit/inch this leads to surface densities of 9–50 Gbit/inch[2]. Given the current track densities, the track pitch is in the range of 250 nm (100 000 tracks/inch) to 850 nm (30 000 tracks/inch). Controlling the head position to a precision of 10% of the track width, to make sure that most of the track width adds to the signal and to keep noise at bay, results in a required precision of 25 nm–85 nm when controlling the track following.

7.5.3   Circuit development for disk drives

Designing circuits for disk drives is currently an ASIC[4] business with a low number of customers, i.e. the disk drive companies, and high volumes. For instance, 223 million disk drives were shipped in the year 2000. This is still the case even though several attempts to create ASSPs[5] have been made in the past and are underway at the moment. The electronics of a disk drive — together with the related firmware — can be partitioned into a few functions:

- **Servo control**

  The servo control detects the current position of the head using so-called servo marks which are embedded into the tracks. Several hundreds of these marks are available throughout a full rotation of the disk. On the basis of this information the voice coil motor is controlled to allow for track following and seeking.

---

[3] Note that none of the dimensions of a 3.5 inch drive is actually 3.5 inch. The disk's form factor is just the same as for a 3.5 inch floppy drive.
[4] Application specific integrated circuit.
[5] Application specific standard product.

- **Spindle control**

  The spindle control detects the current rotational velocity, e.g. by means of back-emf[6] evaluation of the undriven phase of the brushless DC spindle motor. This information is taken into account in the control of the rotational velocity of spindle and disk stack. Spindle control also includes the (electronically controlled) commutation of the motor's phases and the implementation of sophisticated start-up algorithms. The problem with the start-up is that the brushless DC motor has no preferred rotation direction. However, if the disk rotates in the wrong direction the read/write-head might easily damage the disk's surface. So a great deal of work is underway into the sequence of the spindle motor start-up commutation to reliably provide for the correct rotation direction.

- **Signal processing**

  The signal processing function controls the data as it is written to and read from the disk. It encodes the data to be stored, and writes it to the read/write-head. When reading from the read/write-head, the data is pre-amplified and retrieved, using sophisticated techniques like the partial response maximum likelihood method. To allow for even more density — i.e. an even worse signal to noise ratio — the data on the disk contains additional information which is used for error correction. This often is based on Reed/Solomon techniques.

- **Buffer management**

  Data written to or read from the disk has to be fed through a buffer of substantial size to reliably rule out over- and under-runs. For this, DRAM of typically 0.5–2 MBytes is employed, which may be stand-alone or embedded in a system on a chip. In addition, a buffer manager is necessary that offers caching facilities and manages access to the memory for data buffering and other purposes, e.g. program execution from the DRAM.

- **Host interface**

  The host interface implements the communication with the host computer. It interprets the IDE or SCSI commands given and arranges their completion.

In the design process these functions are mapped to an architecture. The list of functions and the overall architecture is more or less generic and is the same for most of the disk drive designs, see Figure 7.14. The overall architecture of the disk electronics is based on five circuits, see Figure 7.14: motion-control, hard disk controller (HDC), RW-channel (stand-alone or embedded), pre-amp and DRAM (stand-alone or embedded). The HDC provides the micro-controller and also looks after the host interface and — together with the DRAM — the buffer management. The RW-Channel — together with the pre-amp — incorporates the signal processing function. Finally motion-control looks after the analogue and power side of servo

---

[6] Electro-motive force.

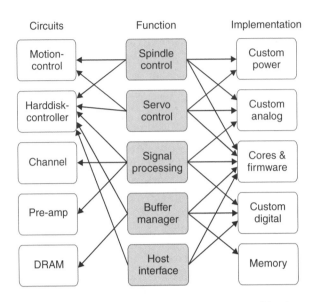

Figure 7.14   Typical mapping of functions to circuits and implementations

and spindle control. When it comes to the detailed architecture, a couple of trade-offs, e.g. performance (hardware) vs. flexibility (software), lead to a vast number of variants in use. For instance, for the control-related parts of the disk function, one may choose to use a standard micro-controller with some DSP-capabilities, e.g. multiply-accumulate operation. If performance plays a more important role, a DSP core may be added. Moreover, in some cases it makes sense to introduce application specific logic for the control tasks.

For the implementation, the classical options are available, i.e. processor cores with firmware, custom digital circuitry, custom analogue circuitry, custom power circuitry and memory. Figure 7.14 shows the mapping of the functions into their implementation.

Under the system-on-chip regime, the above five circuits may be arranged on three to five chips. The pre-amp will probably remain in the form of a stand-alone device in the future, since it is located on the load-beam, i.e. the shortest possible distance from the read/write-head. All other circuits are on the printed circuit board which is on the reverse of the disk drive. The motion-control circuit contains a couple of power devices for which integration into the digital standard process would not be cost-effective. Thus it probably will not be embedded into current or future SoCs.[7] On the other hand, the hard disk controller may be combined with an embedded channel and embedded DRAM on a chip to reduce package cost, save board space and increase the electrical performance. Figure 7.15 shows a couple of alternatives for the mapping of the five circuits to chips.

---

[7] System on a chip.

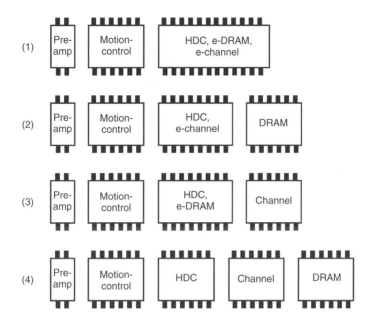

Figure 7.15 Several alternatives for the mapping of circuits to chips

7.5.4 The virtual disk drive

The design of circuits for disk drives has to take into account analogue–digital, hardware–software and electronics–mechanics interfaces at the same time. The Virtual Disk Drive was created to cope with that. It provides simulation models for all the parts of a disk drive and is meant to support an electronics product for a disk drive throughout its entire life-cycle.

- In the concept phase, high-level models form an executable specification which can be validated by simulation. This model also serves as a framework for design space exploration and even allows for early firmware development.

- In the design phase, the virtual prototype creates a general test bench. In this high-level system model, it is possible to 'zoom' into the components being designed, replacing high-level component models by their implementations. This may be done for any part of the system, whereas the system modelling has to be carried out just once. Moreover, this reveals information on the real-life system behaviour rather than more or less synthetic data on signals at a component's interface. For a disk drive the focus may, for instance, be on spin-up, track-following or seeking.

- The development of a high-level (sub)system model is also indispensable for the virtual testing of the analogue circuitry. The test program development may be carried out much earlier in combination with a model of the testing equipment.

- Finally, when the product is with the customer, i.e. a disk drive company, or even in the field, the analysis of spurious behaviour on the system model is much easier for application engineers, since every signal or quantity is visible. Unfortunately, this does not cover all possible faults of a device, since implementation related effects are not taken into account in system modelling. On the other hand, with adequate fault modelling it should be easy to prove or disprove any hypothesis that the application engineers may have on the root cause of system failures.

In the following, this will be illustrated on the basis of the example of the servo control of a disk drive system.

7.5.5 System modelling

A disk drive comprises digital and analogue/power electronics, firmware and mechanics which — as pointed out before — results in three major interfaces: hardware–software, analogue–digital and electronics–mechanics. These interfaces have to be handled at the same time, which does not necessarily mean that the complete drive has to be present in any system simulation, but for instance when assessing the servo control for controlling the track following or seeking, all interfaces are present, see Figure 7.16. Without loss of generality, we restrict ourselves to the servo control in the following.

The model comprises the seeking and track-following controller which is formulated in C and in the real system is implemented in firmware. Its inputs are the current track and the required track. Its output is a digital value for the required current to drive the head assembly. According to this value, the current is regulated

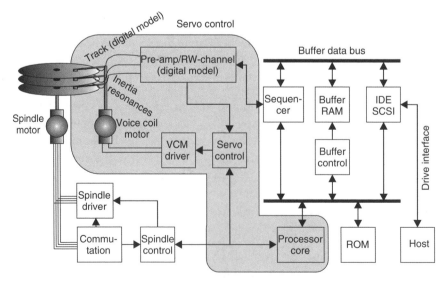

Figure 7.16 Disk drive schematic with servo control

by an analogue/power circuit, i.e. motion-control. The voice coil motor transforms the current into motion of the head assembly which may or may not contain resonances. The mechanical motion results in a track position which is fed into the firmware controller. The track content including track-id — as well as the logic to detect it — is reduced to a more or less trivial digital model, since it does not add to the overall function of the servo control.

The simulation of the servo control is performed on the basis of the mixed-mode simulator Saber. The analogue part of the system — motion-control and mechanics — are modelled in the analogue hardware description language MAST. Some basic digital modelling including the synchronisation between the regulator C-routine and the rest of the system is carried out using the digital capabilities of MAST.

7.5.6 Simulation and results

The high-level model of the servo control as described in the previous section can be used for concept engineering. For example, it is trivial to change the number of servo fields in a track, which at a fixed rotational speed determines the frequency of position measurements and thus the frequency of the digital controller. In the same manner, most of the other variables determining the function and performance of a drive may be varied. This can even be carried out systematically, e.g. by a parameter sweep or — if more than one parameter is involved — on a Monte-Carlo basis. All these variations can be simulated efficiently. For instance, the simulation of a long seek over 20 000 tracks takes about 80 CPU seconds for 20 ms real-time on a SUN Ultra 60 workstation, see Figure 7.17. One can easily spot the 'bang-bang' strategy of maximum acceleration, constant speed at the speed-limit and maximum deceleration to lock into the new track position. Incidentally, the speed limit is due to the fact that the read/write-head rides on an air cushion. This requires an air-stream in the direction of the tracks, which is created by the rotations of the disk(s). The movement for track seeking is perpendicular to that and could seriously disturb the mechanism described above beyond a certain speed. In the next step, the motion-control part of the servo control model was replaced by its implementation, represented by about 300 CMOS transistors and some DMOS power transistors. The real-life circuitry was thereby verified in The Virtual Disk Drive, which is far more meaningful than the results of classical analogue test benches. With the same configuration as the previous long-seek analysis, the simulation takes about 9 CPU hours. Note that most seek operations are performed much quicker and that track-following can be reasonably assessed in an even shorter period. In addition to the standard tasks of seeking and track following, it is now easily possible to review the implementation of special features, e.g. a request to park the heads in the landing zone.

The virtual testing of analogue circuitry is currently under evaluation. The development of a high-level system model is absolutely indispensable to this. This will

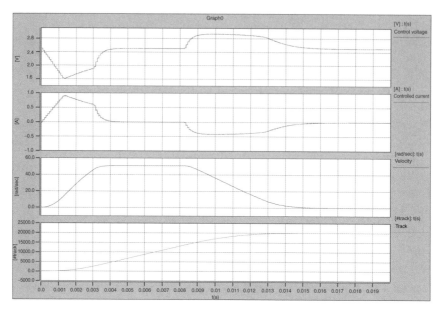

Figure 7.17   Simulation results for long seek over 20 000 tracks. From top to bottom: digital output of firmware controller, controlled current, velocity of rotational actuator and track-number

be combined with a model of the test equipment, which allows test program development to be initiated long before silicon is available.

The support of application engineering is mostly about analyzing the system to explain spurious behaviour, which in turn allows solutions to be devised and implemented. This includes fault modelling for the components of any domain. One might easily include, amongst many others:

- Head assembly resonances of certain frequencies (mechanics)

- On-resistances of power transistors higher than specified (analogue electronics)

- Timing problems in the serial communication between motion-control and disk controller (digital electronics)

This facilitates the assessment of the influence of the respective fault on system behaviour, which in turn can be compared to the behaviour observed in field.

7.5.7 Conclusion

The section has given an overview on state of the art disk drive technology and indicates how this translates into requirements for disk electronics. Moreover, it has shown how system modelling and simulation can support a related electronics product throughout its entire life-cycle. In that vein, the multi-domain nature of the system increases both the need for a systematic solution and the problems associated with modelling and simulation at the same time.

7.5.8 Acknowledgement

The author thanks his colleagues, Wolfgang Sereinig, Wolfgang Sauer, Felix Markhovsky and Scott Burnett for their valuable input and support.

7.6 Summary

Various examples of mechatronic systems have been presented in this chapter. These include electronics, mechanics and sometimes also software. In all cases a complete modelling could be achieved on the basis of hardware description languages, which means that the system in question can be investigated using a simulator.

8

Micromechatronics

8.1 Modelling Micromechatronic Systems

8.1.1 Introduction

Micromechatronic or microelectromechanical systems (MEMS) represent a significant part of microsystem technologies. We will consider surface micromechanics as an example here, because in this field it is often possible to design the process steps for micromechanics to be compatible with those for microelectronics, which permits an integration of electronics and mechanics on a die. The dynamic interactions between electronics and mechanics often cannot be disregarded in such systems. Rather, the desired functionality is often only achieved by the close coupling between the domains. Furthermore, the natural frequencies of the mechanical and electronic oscillations often lie in the same range. This tends to be rare for the general case of an electro-mechanical system and is primarily associated with small mechanical component dimensions in the range of a few tens to a few hundreds of microns. With typical material parameters this yields mechanical resonance frequencies in the megahertz range. Thus a coupled simulation is a crucial prerequisite for the design of electro-mechanical microsystems.

A glance at the development of microelectronics over the last 20 years helps to characterise the situation. The availability of efficient simulation tools was a key factor for dynamic development of the integrated circuits. Precise and efficient models of basic components were, and remain, vital prerequisites for this. In the case of microelectronics only very few components, such as MOS transistors, diodes, capacitors or resistors, have to be taken into account. The introduction of micromechanics has changed matters. For MEMS a variety of nonelectrical components, with a still much greater number of geometric variations, have to be modelled, which leads to significant problems.

The modelling and simulation of MEMS can be considered on component or system level. In the first case, the emphasis is placed upon the design and optimisation of micromechanics, the second case focuses on how a micromechanical component behaves in the context of the system, i.e. within a circuit, for example,

Mechatronic Systems   Georg Pelz
© 2003 John Wiley & Sons, Ltd   ISBN: 0-470-84979-7

and how the entire system can be designed and optimised. The methods of the first category often form the basis for the consideration of the higher abstraction level.

8.1.2 Component design

A whole range of CAE methods have been proposed for the design of microme-chanical components in recent years, see for example Funk *et al.* [108], Puers *et al.* [341], Sandmaier *et al.* [358] or Zhang *et al.* [436]. The MEMCAD system, described by Senturia *et al.* [381] provides a good example of this. Figure 8.1 shows an overview of the structure of this CAE system. In the first stage a so-called structure simulator is used in order to depict the descriptions of the mask layout and the process sequence provided by the user in the three-dimensional geometry of micromechanics, along with a process history for the selection of the appropriate materials. This is facilitated by the repeated call up of a lower-level process simulator. Whilst the geometry becomes a 3D model by meshing in finite elements or boundary elements, we can use the information of the process history to select the appropriate material parameters in the database. This is combined with the 3D model and can then be subjected to a mechanical or electrostatic analysis. Often such analyses should be coupled together, because the underlying phenom-ena, such as, for example, elasticity mechanics, inertia mechanics, electrostatics, or fluidics,[1] should be considered jointly. Two fundamental approaches have been

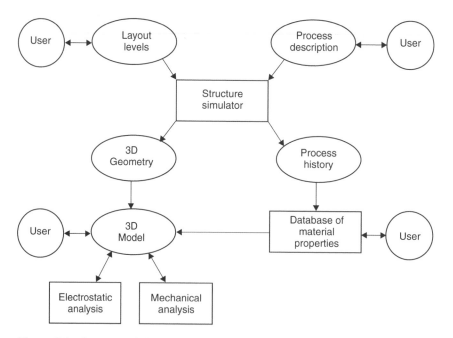

Figure 8.1   Structure of the MEMCAD system, see [381], for the design of MEMS

---

[1] For the description of damping of the mechanical movement in gases and fluids.

developed in recent years for the solution of this problem. Firstly, we can consider the various domains using coupled FE or BE simulators,[2] see, for example, Cai *et al.* [57], Gilbert *et al.* [119] or Senturia *et al.* [381]. The other possibility is to accommodate the various domains in *one* FE simulator on the basis of appropriate finite elements, such as, for example, in Funk *et al.* [108]. In the first case the participating simulators are called up sequentially until a state is reached that is both stable and also compatible with all domains. Such a state is also called self-consistent. In the other case, the analyses of the various domains within a simulator are iterated, with a self-consistent solution only being reached after a while.

In addition to the analysis of the predetermined design, the exploration of the design space is also supported in some cases. For example, Lee *et al.* [223] propose a technique called 'Design-Window', which replaces the single passage through all predetermined parameter combinations with a search led by a neural network. The number of iterations through automatic 3D modelling, meshing, and FE simulation operations run by the system, is thereby minimised.

8.1.3 System design

In contrast to the previous section, in system design the environment of the components has to be taken into account. In our context this is normally a circuit that either effects the triggering of a micromechanical actuator or the read-out of a micromechanical sensor.

Which models are suitable for describing the mechanics in a MEMS? A whole range of criteria can be drawn up here, which should ideally be fulfilled. In most cases, however, we are still a long way removed from this. In our context a mechanics model should:

- be sufficiently precise to correctly represent reality;

- be efficiently simulatable, so that the computing time remains within reasonable limits;

- not cause numerical problems;

- permit the setting of all significant design and technology parameters, in order to thus ensure the general applicability of the model;

- describe (quasi-)static and dynamic behaviour;

- be able to formulate the retention and dissipation of energy in relation to the application.

In addition, there is also the problem of determining the main material and technological parameters. Examples of these parameters are fabrication-related

---

[2] FE: finite elements, BE: boundary elements.

compressive or tensile stresses, the modulus of elasticity, or the dielectric constants. In some cases no values can be found in the literature. For example, the modulus of elasticity of polycrystalline silicon depends, amongst other things, upon its grain structure and its doping, and thus has to be determined individually for each case. This typically occurs on the basis of test structures, which are used for parameter extraction and are incorporated into process control, as in microelectronics, see for example Kiesewetter *et al.* [192] or Voigt *et al.* [415]. If no precise values can be provided here, the reliability of the simulation is jeopardised.

The two demonstrators represented in the following describe the structural modelling for a circuit simulator based upon finite elements. The first example relates to a capacitive pressure sensor in surface micromechanics, see Dudaicevs *et al.* [87] and [88]. The second example is an actuator in the form of a micromirror, which is arranged in an array on a chip, can be deflected individually, and in this manner generates pixel images for all types of displays, see Kück *et al.* [209], Younse [433] and Bielefeld *et al.* [29].

8.2 Demonstrator 5: Capacitive Pressure Sensor

8.2.1 System description

Using the approach described above, microsystems are investigated that cannot be described simply by the mass/spring/damper paradigms, i.e. nonsuspended systems. A good example for this is the integrated, capacitive pressure sensor system in surface micromechanics. Figure 8.2 shows a chip photo of such a system.

The fundamental principle[3] is based upon the fact that an external liquid or gas pressure bends the upper plate of the sensor element downwards, see Figure 8.3 and Figure 8.4. This leads to a change in the capacitance, which is detected by a read-out circuit. The system behaviour depends upon the elastic line of the plate, because it determines the capacitance of the arrangement. Consequently, the precise strain should be taken into account during an electro-mechanic simulation. In addition, the fact that the determination of the capacitance requires the application of a read-out voltage, which itself causes an electrostatic force between the plates, also gives rise to parasitic feedback. This again changes the deflection, which leads to a variation of the capacitance read.

The conversion of the capacitance into an output voltage is carried out in two stages using 'switched capacitor' technology. The first stage compares the sensor capacitance with a passivated reference capacitance. As a result of this differential processing, the result is barely influenced by manufacturing-related deviations. The second stage is the amplification of the read-out signal. Finally the output signal is smoothed by a 'sample and hold' stage. The circuit contains around 700 analogue components. The eighteen sensor and reference elements are each connected in

---

[3] See also the description in Section 6.3.3.

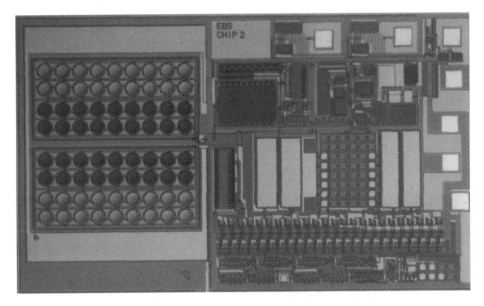

Figure 8.2 Chip photo of the integrated, capacitive pressure sensor system in surface microme-chanics (Reproduced by Permission of Gerhard-Mercator-University, Duisburg, Germany)

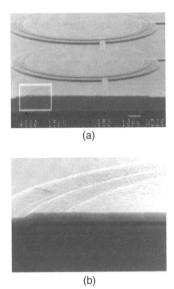

Figure 8.3 SEM photo of the cross-section of a pressure element (Reproduced by Permission of Fraunhofer-Institute IMS, Duisburg, Germany)

parallel. This array of eighteen active and eighteen passivated pressure elements, is duplicated in order to achieve still higher precision using differential path tech-nology. For each set of eighteen pressure elements, a concentrated component is used, the capacitance of which is correspondingly multiplied.

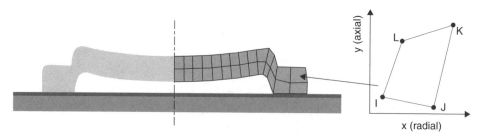

Figure 8.4 Arrangement of finite elements for the modelling of the micromechanical pressure element (left); plane element each with the degrees of freedom u and v per node (right)

The described system consists primarily of two sections: the conversion of pressure into capacitance and the read-out of the capacitance. Special emphasis lies on the investigation of the system behaviour. This means that we apply a pressure profile and consider the resulting voltage profile. In addition to the actual functionality, however, further properties of the sensor have to be investigated; for example, its sensitivity or linearity.

8.2.2 Modelling

Introduction

The circuit consists of the following components: 634 MOS field effect transistors, 52 resistors, 19 capacitors and 12 diodes. These components are either modelled in hardware description languages or permanently integrated into the simulator. We can generally rely upon the component models offered by the simulator here.

The modelling of the mechanics is more difficult, see also Pelz *et al.* [333] and Bielefeld *et al.* [33]. Two fields of 18 pressure elements each—one is passivated, the other is non-passivated—have to be considered. For modelling it is sufficient to describe one element each time and to multiply the capacitance values by 18. The passivated boxes can be simply replaced by a reference capacitance here because their capacitance deviation can be disregarded in this case.

The actual modelling of the pressure elements takes place by a radial section, which is justified by its rotationally symmetrical structure, see Figure 8.4. The section is described by a number of finite (plane) elements. In the simplest investigated case 28 elements are used, as illustrated. Each finite element has four nodes, each with two degrees of freedom, u and v, the deflection in the x and y directions. Multiplying the number of elements and the number of degrees of freedom of an element results in a total of 224 degrees of freedom. The actual number of degrees of freedom is significantly less, i.e. only 81. This is because many nodes of adjacent elements lie on one another, so that the degrees of freedom in question coincide.

The interpolation functions, and thus the element matrices, for the implementation of the plane element correspond with the solution selected in the FE simulator

ANSYS. Thus, as shown later in the simulation experiments, the mechanical behaviour cannot be distinguished from that simulated using ANSYS.

Setting up the element matrices

Now, the finite element to be formulated in a hardware description language should be parametrisable with regard to its dimensions and the associated material constants. Furthermore, the model should also be suitable for greater deflections. Both cases require that the element matrices must be constructed in the model of the finite elements. In this way the approach described here, see [333] and [33], differs significantly from the (later) work of Haase *et al.* [131] in which a system matrix of the mechanics constructed using a FE simulator is imported into a circuit simulator via a hardware description language.

The construction of the element matrices should be represented for an isoparametric, rotationally symmetrical plane element, see also Bathe [19]. For this purpose a natural coordinate system is introduced for the element, which has the coordinates s and t, which each run from -1 to $+1$ see Figure 8.5.

The first step is to approximate the coordinates of arbitrary points of the plane elements on the basis of the coordinates of the four nodes. To this end the form functions h_i are introduced, which are formulated as follows in the natural reference system of the element:

$$h_1 = \tfrac{1}{4}(1-s)(1-t) \qquad h_2 = \tfrac{1}{4}(1+s)(1-t)$$
$$h_3 = \tfrac{1}{4}(1+s)(1+t) \qquad h_4 = \tfrac{1}{4}(1-s)(1+t) \qquad (8.1)$$

The fundamental property of these functions is that the h_i becomes equal to 1 at node i and becomes equal to 0 at all other nodes.

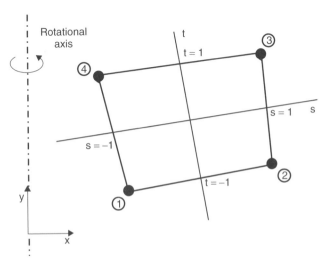

Figure 8.5 Natural coordinate system for a rotationally symmetrical plane element

The local coordinates of arbitrary points (x, y) of the element are thus found to be:

$$x = \sum_{i=1}^{4} h_i x_i, \qquad y = \sum_{i=1}^{4} h_i y_i \tag{8.2}$$

where x_i or y_i denote the coordinates of the element nodes. In the isoparametric formulation the element displacements are interpolated similarly to the coordinates, but this time the node displacements are multiplied by the form functions.

$$u = \sum_{i=1}^{4} h_i u_i, \qquad v = \sum_{i=1}^{4} h_i v_i \tag{8.3}$$

Here u and v are the local element displacements at an arbitrary point of the element and u_i and v_i are the corresponding element displacements at the nodes. The node displacements are furthermore summarised in the form of a vector:

$$\hat{u}^T = \begin{bmatrix} u_1 & v_1 & u_2 & v_2 & u_3 & v_3 & u_4 & v_4 \end{bmatrix} \tag{8.4}$$

Thus (8.3) can be drawn up in matrix form:

$$u(s, t) = H\hat{u} \tag{8.5}$$

The next step is the transition from the element displacements to the element strain ε, which is defined by the derivative of the displacements with respect to the local coordinates:

$$\varepsilon^T = \begin{bmatrix} \varepsilon_{xx} & \varepsilon_{yy} & \gamma_{xy} & \varepsilon_{zz} \end{bmatrix} = \begin{bmatrix} \dfrac{\partial u}{\partial x} & \dfrac{\partial v}{\partial y} & \dfrac{\partial u}{\partial y} + \dfrac{\partial v}{\partial x} & \dfrac{\partial w}{\partial z} \end{bmatrix} \tag{8.6}$$

The last entry of ε corresponds with the hoop strain, which is triggered by a displacement in the x-direction over the whole circumference according to the rotational symmetry.

We can thus draw up the strain-displacement transformation matrix **B**, which effects the transformation of the displacements to the strains.

$$\varepsilon = B \cdot \hat{u}, \qquad B = \begin{bmatrix} \dfrac{\partial h_1}{\partial x} & 0 & \dfrac{\partial h_2}{\partial x} & 0 & \dfrac{\partial h_3}{\partial x} & 0 & \dfrac{\partial h_4}{\partial x} & 0 \\[2mm] 0 & \dfrac{\partial h_1}{\partial y} & 0 & \dfrac{\partial h_2}{\partial y} & 0 & \dfrac{\partial h_3}{\partial y} & 0 & \dfrac{\partial h_4}{\partial y} \\[2mm] \dfrac{\partial h_1}{\partial y} & \dfrac{\partial h_1}{\partial x} & \dfrac{\partial h_2}{\partial y} & \dfrac{\partial h_2}{\partial x} & \dfrac{\partial h_3}{\partial y} & \dfrac{\partial h_3}{\partial x} & \dfrac{\partial h_4}{\partial y} & \dfrac{\partial h_4}{\partial x} \\[2mm] \dfrac{h_1}{R} & 0 & \dfrac{h_2}{R} & 0 & \dfrac{h_3}{R} & 0 & \dfrac{h_4}{R} & 0 \end{bmatrix} \tag{8.7}$$

where R is the radius relating to the rotation. The hoop strain is calculated as:

$$\frac{\partial w}{\partial z} = \frac{u}{R} \tag{8.8}$$

The element displacements u and v are, however, defined in the natural coordinate system (s,t), so that the derivatives with respect to x and y must be linked to the derivatives with respect to s and t. According to the chain rule, the following applies:

$$\begin{bmatrix} \partial/\partial s \\ \partial/\partial t \end{bmatrix} = \mathbf{J} \begin{bmatrix} \partial/\partial x \\ \partial/\partial y \end{bmatrix} \quad \text{where} \quad \mathbf{J} = \begin{bmatrix} \partial x/\partial s & \partial y/\partial s \\ \partial x/\partial t & \partial y/\partial t \end{bmatrix} \tag{8.9}$$

It is a prerequisite that a clear relationship between the coordinate systems exists. In this case the following applies:

$$\begin{bmatrix} \partial/\partial x \\ \partial/\partial y \end{bmatrix} = \mathbf{J}^{-1} \begin{bmatrix} \partial/\partial s \\ \partial/\partial t \end{bmatrix} \tag{8.10}$$

By the use of the inverted Jacobi's operator $\mathbf{J}^{-1}$ we can now set up the strain vector for arbitrary points on the rotationally symmetrical plane elements.

We then move from the strains $\boldsymbol{\varepsilon}$ to the stresses $\boldsymbol{\tau}$, by using the material matrix $\mathbf{C}$:

$$\boldsymbol{\tau} = \mathbf{C} \cdot \boldsymbol{\varepsilon} \quad \text{where} \quad \boldsymbol{\tau}^{\mathrm{T}} = [\tau_{xx} \, \tau_{yy} \, \tau_{xy} \, \tau_{zz}] \tag{8.11}$$

For the material matrix of a rotationally symmetrical element the following applies according to [19]:

$$\mathbf{C} = \frac{E(1-v)}{(1+v)(1-2v)} \begin{bmatrix} 1 & \dfrac{v}{1-v} & 0 & \dfrac{v}{1-v} \\ \dfrac{v}{1-v} & 1 & 0 & \dfrac{v}{1-v} \\ 0 & 0 & \dfrac{1-2v}{2(1-v)} & 0 \\ \dfrac{v}{1-v} & \dfrac{v}{1-v} & 0 & 1 \end{bmatrix} \tag{8.12}$$

In the material matrix, E represents the modulus of elasticity and v represents Poisson's ratio. To determine the stiffness matrix we first apply the elastic potential of a plane element:

$$\Pi = \frac{1}{2} \int_A \boldsymbol{\varepsilon}^{\mathrm{T}} \boldsymbol{\tau} \, dA \tag{8.13}$$

Alternatively we can also formulate

$$\Pi = \frac{1}{2} \hat{\mathbf{u}}^{\mathrm{T}} \mathbf{K} \hat{\mathbf{u}} \tag{8.14}$$

If we finally equate (8.13) and (8.14), then using (8.7) and (8.11) we find the following expression for the stiffness matrix:

$$K = \int_A B^T C B \, dA \tag{8.15}$$

Because the coordinates of the matrix B are defined in the natural coordinates, the integration must be performed using these coordinates

$$dA = \det J \, ds \, dt \tag{8.16}$$

Substituting yields:

$$K = \int_A F \, ds \, dt \quad \text{where} \quad F = B^T C B \det J \tag{8.17}$$

Since the analytical integration is difficult to get to grips with, at this point a numerical integration will be performed on the basis of the Gauss–Legendre quadrature. To this end the following support points are used in natural coordinates:

Support points (i,j)	s_i	t_j	α_{ij}
(1,1)	−0.577350269189626	−0.577350269189626	1.0
(1,2)	−0.577350269189626	+0.577350269189626	1.0
(2,1)	+0.577350269189626	−0.577350269189626	1.0
(2,2)	+0.577350269189626	+0.577350269189626	1.0

For every support point the matrix F_{ij} has to be evaluated and multiplied by the factor α_{ij}. The result is summed and forms the element stiffness matrix:

$$K = \sum_{i,j} 2\pi R_{ij} \alpha_{ij} F_{ij} \tag{8.18}$$

Here F_{ij} of the matrix corresponds with F at the integration points s_i and t_j. The values of α_{ij} are weighting factors that are determined for the numerical integration. Finally, the factor $2\pi R_{ij}$ represents the circumference with regard to the rotation at the integration point (s_i, t_j) and thus the 'thickness' of the element.

The creation of the element mass matrix is similarly completed in accordance with the equation:

$$M = \int_A \rho H^T H \, dA \tag{8.19}$$

where ρ represents the material density and H the transformation matrix from (8.5). The above-mentioned operations are implemented in the programming language C.

This is based upon a library of matrix operations. As the following section shows, the C routines for element stiffness matrix and element mass matrix are called up from the hardware description, in order to determine the matrices in question. The above-mentioned operations for the creation of the element matrices should be performed at least once at the beginning of the simulation and several times in the event of greater deflections or nonlinearities.

Formulation in a hardware description language

The finite plane element called `plane_u2` described in the previous sections will now be formulated in an analogue hardware description language, see Hardware description 8.1. The MAST language has been selected for this. Two aspects have to be taken care of: the creation of the mass and stiffness matrices of the elements dealt with in the previous section and the linking of the mechanics thus described into a circuit simulation, see Chapter 6.

First of all the question of the terminals and parameters of the model arises. With its eight degrees of freedom the plane element should also possess eight terminals, which each represent one degree of freedom. These are deflections of the nodes i, j, k and l in the x and y directions, so the pins will be called `uxi`, `uyi`, `uxj`, `uyj`, `uxk`, `uyk`, `uxl` and `uyl`. With regard to the parameters, we differentiate between geometric and material parameters. The former specify the geometry of the element, i.e. the position of the nodes i, j, k and l in nondistorted state, and are called `xi`, `yi`, `xj`, `yj`, `xk`, `yk`, `xl` and `yl`. The latter are the density (`dens`), the modulus of elasticity in the x direction (`ex`) and y direction (`ey`), the shear modulus (`gxy`) Poisson's ratio (`nuxy`) and information about whether the element is to be considered as rotationally symmetrical (`plstr`).

There follows the declarations of the C routine (`foreign`), which calculates the element matrices, declarations of the element entries themselves (`l1_1`, `c1_1 ...`) and the introduction of the branch (`branch`) that associates the value of internal variables with voltages and currents. In the following Values section the aforementioned C routine is called up in order to determine the element matrices. Depending upon the type of the finite element and application case, this may take place once at the beginning or as required during the simulation. Finally, the equation section includes the equation system that describes the element, in which each equation defines capacitive or inductive behaviour.

```
template plane_u2 uxi uyi uxj uyj     \
                  uxk uyk uxl uyl =   \
                  xi, yi, xj, yj,     \
                  xk, yk, xl, yl,     \
                  dens, ex, ey, gxy,  \
                  nuxy, plstr
   electrical uxi,  # x-deflection, node i
              uyi,  # y-deflection, node i
      . . . . .
```

```
   number xi = 0.0, # x-coordinate, node i
          yi = 0.0, # y-coordinate, node i
      . . . . .
   number dens = 0.0, # Density
      . . . . .
{
   foreign planeu2     # C function
      . . . .
   val nu l1_2,l1_3,l1_4, # L coefficient
      . . . .
          c1_1,c1_3,       # C coefficient
      . . . .
   # Definition of branch voltages and currents
   # at node uxi
   branch i1_1a = i(uxi->gnd), \
          v1_1 = v(uxi,gnd)
   . . . .
   . . . .
   # at node uyl
   branch i8_8a = i(uyl->gnd), \
          v8_8  = v(uyl,gnd)
   # Call up of the external C function for the calculation
   # of the mass and stiffness matrix of the element
   values{
      . . .
      (l1_2,l1_3,. . . . . .     \
       c1_1,c1_3,. . . . .) =  \
      plane_u2 (xi, yi, . . . . .)
      . . .
   }
   # Definition of the dynamic equations
   equations{
      i1_1a = d_by_dt(v1_1  * c1_1)
      v1_2  = d_by_dt(i1_2b * l1_2)
      . . . .
      i8_8a = d_by_dt(v8_8  * c8_8)
   }
}
```

Hardware description 8.1   Description of the finite plane elements, each with four nodes and two degrees of freedom

System matrix

Using the element description obtained in this manner the deflections of the finite elements in question are represented by the voltages at the terminals of the element. The currents at the nodes in question describe the integral of the associated forces and moments, depending upon whether the degree of freedom relates to a translational or rotational deflection. In particular, the parts of the exciting forces and moments that are assigned to the elements adjoining the node are also added to the currents (and current changes) at a node. It is not necessary to explicitly create the system matrix, since its solution is yielded implicitly from the interconnection of the finite elements.

Excitation of the mechanics

In addition to the mechanical behaviour of the structure it is necessary to describe the external mechanical excitation. For reasons of modularity this is formulated in a further element, which is connected to the finite plane elements in question at the node(s) where the excitation acts. The main task of this element is to determine the exciting forces and moments at each node point, i.e. for each degree of freedom, and to convert these into a current change, with the resulting current being fed into the circuit nodes of the degree of freedom in question. The same procedure can in principle also be used in order to model the electrostatic feedback using an additional element. However, this effect could mainly be disregarded in the application considered, so that the corresponding modelling could be dispensed with.

Geometric nonlinearity

If the deflection exceeds a certain value, then the upper plate rests upon the isolator. If the pressure is further increased the plate 'rolls' out over the isolator. This corresponds with a one-sided holonomous and scleronomic constraint of the movement. Numerically this means that upon the manifestation of new boundary conditions, a corresponding number of degrees of freedom disappear. The realisation of numerical equation solvers with a variable number of unknowns or equations is a difficult task; only very few solvers are set up for this case. For this reason it is worthwhile approximating the prescribed situation by damping the specified degrees of freedom to a significant degree. This again leads to the fact that only minimal movements are possible for these, which in principle corresponds with the desired behaviour. However, this procedure may nevertheless be numerically dangerous because the sudden setting of a high damping — in other words the striking of the plate on the isolator — excites the most important natural frequencies.

Calculation of the capacitance

A further element type is introduced for calculating the capacitance, which just like the plane element or the pressure element is formulated in a rotationally symmetrical manner. For this element the plate capacitor equation is applied to the average deflection of a corresponding annulus. The resulting capacitance, in the form of variable capacitors, is positioned between the two plate potentials and is thus available to the circuit.

8.2.3 Simulation

In the following, various simulations will be performed in order to illustrate the application of finite elements for a circuit simulator. The following cases will be considered here: mechanical deflections, changes of the element capacitance, system simulation and the parametrised simulation of the FE models. A FE model of the sensor structure based upon 48 finite elements was used.

Mechanical behaviour

Initially the mechanical behaviour will be verified on the basis of a comparison with the FE simulator ANSYS. For this purpose three different element radii (120 μm, 70 μm and 50 μm) are predetermined and the deflection of various points on the radius considered, see Figures 8.6, 8.7 and 8.8. The results show virtually no differences between FE and the circuit simulator. This is because precisely the same interpolation functions, and thus element matrices, are used. Thus we lose no precision in relation to FE simulation. Such a simulation on the basis of hardware description languages requires around one CPU minute on a SUN Sparc 20 workstation.

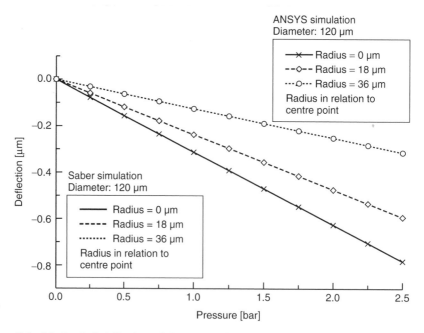

Figure 8.6   Mechanical deflection of the upper plate (diameter 120 μm) at various points on the radius. Comparison of the ANSYS FE simulator with the Saber circuit simulation

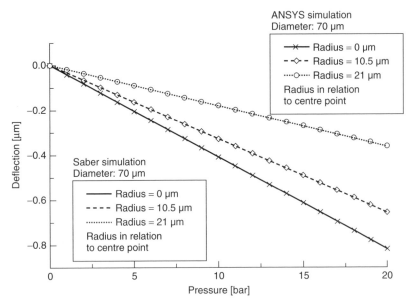

Figure 8.7 Mechanical deflection of the upper plate (diameter 70 μm) at various points on the radius. Comparison of the ANSYS FE simulator with the Saber circuit simulator

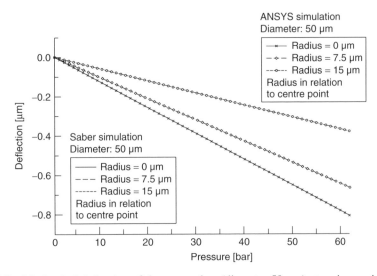

Figure 8.8 Mechanical deflection of the upper plate (diameter 50 μm) at various points on the radius. Comparison of the ANSYS FE simulator with the Saber circuit simulator

The mechanics was very slowly excited in the previous simulations and thus considered as quasi-static. In the modelling, however, the inertia was also specified, so that we can also investigate the kinetics in the simulation. This was taken into account in the following simulation, which shows the reaction of the mechanical structure to a sudden change of pressure, see Figure 8.9.

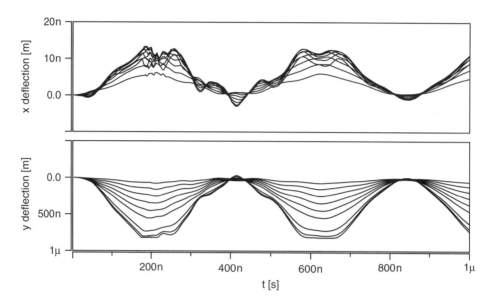

Figure 8.9   Response of the mechanical structure to a sudden change of pressure

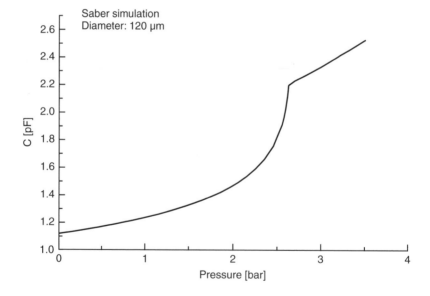

Figure 8.10   Capacitance against pressure for an element diameter of $120\,\mu m$

The x and y displacements of various points on the radius are considered, see
for comparison Figure 8.4. The horizontal (x) displacements are positive and in
the direction of the plate. They are smaller by a factor of 50 than the vertical (y)
displacements, which have negative values and consequently describe a lowering
of the plate. In a vertical deflection of $-800\,\text{nm}$, the plate meets the isolator,

which acts as a mechanical stop and thus causes a geometric nonlinearity. In particular, various natural frequencies are excited as a result of the collision, which is particularly evident from the displacement in the x-direction. It is clear that this type of simulation permits the consideration of the electro-mechanical dynamics in a transient simulation. Furthermore we can read off the most important natural frequencies from such a simulation.

Electrical behaviour

The electrical behaviour of the pressure elements is mainly connected to its capacitance. Figures 8.10, 8.11 and 8.12 show how the capacitance of an element behaves in relation to the pressure. This situation is considered on the basis of a transient simulation, in which the external pressure is increased in a linear manner. We note the kink in the curve, which marks the point at which the upper plate rests upon the isolator. If the pressure increases further the plate 'rolls' over the isolator and exhibits the typical, significantly more linear increase. Furthermore, regarding the various base capacitances at 0 bar, we note that arrays of 6, 18 and 33 pressure elements were selected for the diameters of 120, 70 and 50 μm respectively. The numbers selected correspond with a real fields used for pressure elements.

System simulation

In this section a transient simulation of the pressure sensor system shown in Figure 8.2 will be presented, see Figure 8.13. The applied pressure will be increased

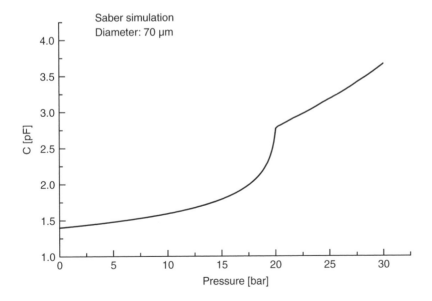

Figure 8.11 Capacitance against pressure for an element diameter of 70 μm

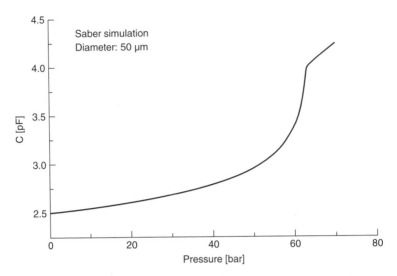

Figure 8.12   Capacitance against pressure for an element diameter of 50 μm

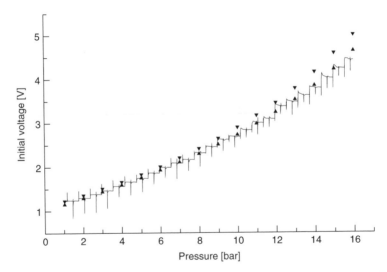

Figure 8.13   Output voltage against pressure for a pressure sensor system. Simulation (solid line), seven series of measurements, the maximum and minimum values of which are each marked by the triangles

continuously from 1 bar to 16 bar. This deflects the pressure element, the change in capacitance of which is converted into an output voltage by the switched capacitor circuit. The simulation (solid line) shows the slightly nonlinear path of measured output voltage. The voltage peaks that occur at regular intervals are caused by the read-out using switched capacitor circuit technology. For the validation of the simulation, seven series of measurements, taken on a corresponding number of manufactured circuits, were considered. The extreme values of these measurements are each depicted by the small triangles. The scatter between the measurements is caused by the scatter between the circuits. The simulation reflects the measured trace with an accuracy of below 10%. It also shows that the simulation remains a little behind the measurements at higher pressures and thus greater deflections. This effect could be caused by the stronger electrostatic forces of the read-out voltage at greater deflections, which are not taken into account in the model. Overall, this type of simulation permits the investigation of the overall function of the system, and also its sensitivity and linearity. This is accomplished even before manufacture. For the case shown, such a system simulation requires around 77 CPU minutes on a SUN Sparc 20.

Parametric simulations

In a circuit simulator it is normal to vary the models based upon parameters, in order to cover as broad an application spectrum as possible. Such parameters can be yielded from the geometry of the structure and from the material properties. A geometric parameterisation, such as the length or width of MOS transistors, can be achieved within certain limits even for FE models. In our case the following parameters were taken into account: diameter of the pressure element; plate thickness and height of the hollow area.

In addition, there are naturally also material properties, for example, the modulus of elasticity, Poisson's ratio or the dielectric constant. A circuit simulator offers the possibility of running certain simulations repeatedly and thus travelling through a predetermined parameter space. This will be performed for one design parameter and one technology parameter, namely for the element diameter and the plate thickness. The prerequisite for this type of simulation is that the geometry of the FE model is variably formulated. This is illustrated in Figure 8.14.

Two parametric simulations are shown in Figure 8.15 and Figure 8.16, in which the diameter and the plate thickness are varied. In this manner the FE model can be used both for design and also for technological optimisation, taking into account the circuit aspects.

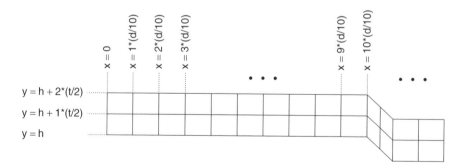

Figure 8.14 Geometric parameterisation of the FE model by diameter d, plate thickness t and height of the hollow space h

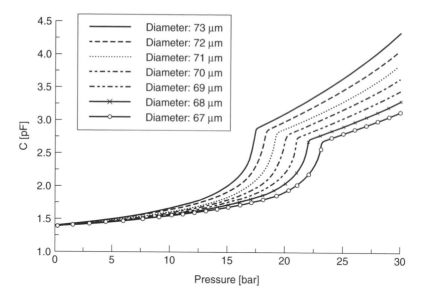

Figure 8.15 Parametric Saber simulations for the variation of the diameter of the pressure element. The plate thickness is fixed at 1.7 μm

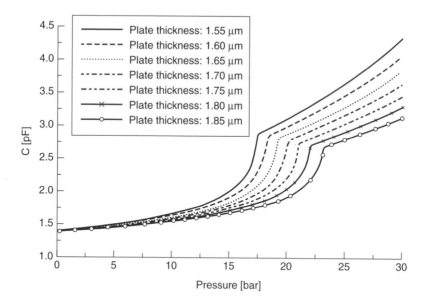

Figure 8.16 Parametric Saber simulations for the variation of the plate thickness of the pressure element. The diameter is fixed at 70 μm

8.3 Demonstrator 6: Micromirror

The second example here is a micromirror, which is arranged in arrays in order to generate pixel images of all types. Just like the pressure sensor described above, this micromirror, see Kück *et al.* [209], is a system manufactured using CMOS compatible surface micromechanics. It is operated as an actuator, i.e. the mechanical displacement is not the object of the measurement, but rather the system behaviour to be caused. Here the deflection consists of the lowering of the mirror. If light falls on the mirror its deflection brings about a corresponding phase shift in the reflected light. The picture to be generated finally arises from the resulting interferences. The deflection is achieved electrostatically.

8.3.1 System description

The micromirror has an edge length of 20 microns and is placed on the chip surface in large arrays, e.g. 512 × 464, see Figure 8.17.

Each mirror is individually addressable and can be moved independently of the others. The mirrors are deflected electrostatically by applying a suitable voltage between the mirror and a counter-electrode located below it. The restoring force of the suspension works against the electrostatic force, so the mirrors return to their initial positions after the voltage is switched off. One problem for modelling is that the resulting force depends significantly upon the distance between mirror and counter-electrode. In particular there is a positive feedback here, which can lead to instabilities.

8.3.2 Modelling

Due to the filigree structure of the micromirror, modelling cannot be achieved analytically without further complications. The finite element method is particularly suited to answering the questions of structural mechanics in such cases. Certain questions are essential to the consideration of the interaction between electronics

Figure 8.17 SEM photo of the micromirror layout (Reproduced by Permission of Fraunhofer-Institut IMS, Duisburg, Germany)

and mechanics. Initially the relationship between the voltage applied and the resulting deflection should of course be investigated. Furthermore, the feedback of the mirror capacitance on the electronics and the possible excitation of resonances of the mirror is also of interest. Figure 8.19 shows the typical deflection of a micromirror and the structure of the associated FE model. This is based upon plate elements, which are particularly well suited for the layer structure of micromechanics. The following representation deals with rectangular plate elements, which are treated in detail in Gasch and Knothe *et al.* [113]. The description of modelling is dealt with more briefly here in comparison with the previous demonstrator because — like the beams introduced in Chapter 6 — small deflections result in constant mass and stiffness matrices for the rectangular plate elements used. Furthermore, the customisation of the element matrices according to geometric and material parameters is very simple in this case.

Each element has four nodes, which lie at the corners of the plate element, see Figure 8.18. Each node again has four degrees of freedom (u_z, r_x, r_y, r_{xy}), where u_z represents the displacement perpendicular to the plane of the plate, r_x and r_y the cross-sectional tiltings in the x and y direction, and r_{xy} the torsion:

$$r_x = -\frac{\partial u_z}{\partial x}, \qquad r_y = -\frac{\partial u_z}{\partial y}, \qquad r_{xy} = -\frac{\partial^2 u_z}{\partial x \partial y} \tag{8.20}$$

The interpolation functions for the rectangular plate element can be obtained in an elegant manner by the multiplication of pairs of interpolation functions of a beam, see for example equation (6.35). One interpolation function covers the x-direction, the other the y-direction. If there are four interpolation functions for a beam there are sixteen interpolation functions for the plate. Using the principle of the virtual displacements we obtain a mass matrix and a stiffness matrix for the plate element, see [113]. The mass matrix depends exclusively upon the density of the material ρ and the dimensions of the plate, see equation (8.21). The stiffness matrix is again influenced by the dimensions, the modulus of elasticity of the material, and Poisson's ratio. In both cases 16×16 matrices are obtained which — as shown

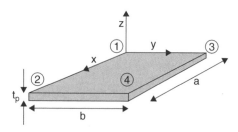

Figure 8.18   Finite element of a rectangular plate with four nodes and four degrees of freedom per node

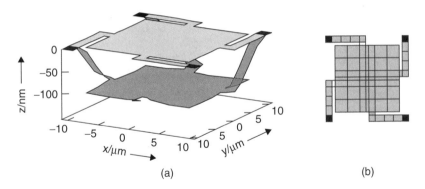

Figure 8.19 Deflection of a micromirror (a), structure of the FE model (b)

in Section 6.3.2–will be represented in hardware description languages.

$$M = \frac{\rho t_p ab}{176400}$$

$$\times \begin{bmatrix}
24336 & 8424 & 8424 & 2916 & -3432 & 2028 & -1188 & 702 & -3432 & -1188 & 2028 & 702 & 484 & -286 & -286 & 169 \\
8424 & 24336 & 2916 & 8424 & -2028 & 3432 & -702 & 1188 & -1188 & -3432 & 702 & 2028 & 286 & -484 & -169 & 286 \\
8424 & 2916 & 24336 & 8424 & -1188 & 702 & -3432 & 2028 & -2028 & -702 & 3432 & 1188 & 286 & -169 & -484 & 286 \\
2916 & 8424 & 8424 & 24336 & -702 & 1188 & -2028 & 3432 & -702 & -2028 & 1188 & 343 & 169 & -286 & -286 & 484 \\
-3432 & -2028 & -1188 & -702 & 624 & -468 & 216 & -162 & 484 & 286 & -286 & -169 & -88 & 66 & 52 & -39 \\
2028 & 3432 & 702 & 1188 & -468 & 624 & -162 & 216 & -286 & -484 & 169 & 286 & 66 & -88 & -39 & 52 \\
-1188 & -702 & -3432 & -2028 & 216 & -162 & 624 & -468 & 286 & 169 & -484 & -286 & -52 & 39 & 88 & -66 \\
702 & 1188 & 2028 & 3432 & -162 & 216 & -468 & 624 & -169 & -286 & 286 & 484 & 39 & -52 & -66 & 88 \\
-3432 & -1188 & -2028 & -702 & 484 & -286 & 286 & -169 & 624 & 216 & -468 & -162 & -88 & 52 & 66 & -39 \\
-1188 & -3432 & -702 & -2028 & 286 & -484 & 169 & -286 & 216 & 624 & -162 & -468 & -52 & 88 & 39 & -66 \\
2028 & 702 & 3432 & 1188 & -286 & 169 & -484 & 286 & -468 & -162 & 624 & 216 & 66 & -39 & -88 & 52 \\
702 & 2028 & 1188 & 3432 & -169 & 286 & -286 & 484 & -162 & -468 & 216 & 624 & 39 & -66 & -52 & 88 \\
484 & 286 & 286 & 169 & -88 & 66 & -52 & 39 & -88 & -52 & 66 & 39 & 16 & -12 & -12 & 9 \\
-286 & -484 & -169 & -286 & 66 & -88 & 39 & -52 & 52 & 88 & -39 & -66 & -12 & 16 & 9 & -12 \\
-286 & -169 & -484 & -286 & 52 & -39 & 88 & -66 & 66 & 39 & -88 & -52 & -12 & 9 & 16 & -12 \\
169 & 286 & 286 & 484 & -39 & 52 & -66 & 88 & -39 & -66 & 52 & 88 & 9 & -12 & -12 & 16
\end{bmatrix}$$

(8.21)

A total of 69 finite elements are used for the modelling of the micromirror, which are connected together as shown on the right-hand side of Figure 8.19. Ninety six nodes have to be considered, with four degrees of freedom each—thus a total of 384 degrees of freedom. In addition to the finite elements of the mechanics, further elements are used to calculate and apply an appropriate force from the applied voltage and the current deflection for each node. Further details with regard to modelling can also be found in Bielefeld *et al.* [34].

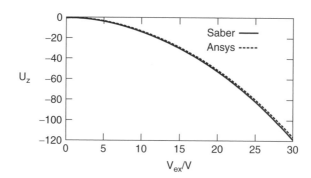

Figure 8.20   Simulation result: deflection of the mirror against the excitation potential; finite elements in hardware description languages (Saber) versus finite elements in a FE tool (ANSYS)

8.3.3   Simulation

In the quasi-static simulation shown in Figure 8.20 the excitation voltage increases linearly from 0 V to 30 V. The deflection of the micromirror is plotted against time or voltage. The results that were achieved using the Saber circuit simulator and the ANSYS FE simulator (at the same discretisation) are shown. The differences between the two simulations are below 2%, so the model formulated in hardware description languages can serve as a good replacement for the FE model. Both simulations require about a minute of CPU time on a SUN Sparc 20 workstation.

8.4   Summary

For virtually all FE simulators it is true that electronic components cannot be included in the simulation to an appreciable degree. So for these tools the consideration of the electronics goes no further than the formulation of electrostatic forces. By contrast, the point of the implementation of the finite elements using hardware description languages is that all models present in the electronics simulator can also be drawn into such a simulation. Thus the development of the electronics can be achieved taking into account continuum mechanics.

This is illustrated by the two demonstrators of this chapter. Both of these are microsystems. Nonetheless, exactly the same methodology is applicable for continuum mechanics of macro scale. We considered a sensor and an actuator, which can both be investigated equally well using the approach.

9

Summary and Outlook

This book is concerned with the problem of modelling and simulating mechatronic and micromechatronic systems on the basis of hardware description languages. To achieve this, it first describes the modelling of electronics, mechanics and electromechanics. This is followed by a comprehensive representation of the description methods and possibilities of hardware description languages. Then the latest methods for the modelling of multibody mechanics, continuum mechanics and software, plus their representation in hardware description languages, are presented. This compendium of basic methods is subsequently applied to six demonstrators from mechatronics and micromechatronics in the application chapters.

In the past, a hotchpotch of isolated individual solutions existed in the field of the modelling and simulation of (micro)mechatronics using hardware description languages. This work makes the transition to a unified approach for the whole problem class by highlighting the most important modelling strategies for the domains typically involved—i.e. multibody and continuum mechanics, digital and analogue electronics plus software—and describing their representation in digital and analogue hardware description languages. The problem of mixed simulation is thereby transformed into a problem of mixed modelling, which is significantly easier to get to grips with. In addition, elaborate, commercial simulators are available for all the main hardware description languages. Furthermore, due to the option of functional modelling, hardware description languages facilitate the introduction of a top-down design sequence. This has brought significant productivity gains in the design of digital electronics and increasingly also in the field of analogue electronics. Using the described methods, top-down design can now also be applied for mechatronic and micromechatronic systems.

In addition to the validation of executable specifications and the verification of designs by simulation—which are already possible using the methods presented in this book—in future, further options will open up, such as the synthesis and formal verification of mixed systems. This used to be the exclusive preserve of digital electronics and to some degree also analogue electronics. The use of the system models available in the form of hardware descriptions removes a significant obstacle.

Mechatronic Systems   Georg Pelz
© 2003 John Wiley & Sons, Ltd   ISBN: 0-470-84979-7

However, there remain a few gaps that will have to be filled in the future. For example, the object-oriented modelling of three-dimensional multibody systems on the basis of hardware description languages is still in its infancy.

Furthermore, a significant problem of software modelling is the support not only of one controller, but of various families of controllers sometimes with some tens of variants. This calls for a set of building blocks that puts the peripheral blocks used together with the processor core to form a model. It will also be necessary to consider various abstractions in order to consider the pure function in one case, but to consider additional parasitic effects in another.

Literature

[1] Abel, H.-B. and Papanuskas, J. (1996), Einsatz von Hardwarebeschreibungssprachen im Entwurf von Kfz-Komponenten, Hardwarebeschreibungssprachen und Modellierungsparadigmen: 2. GI/ITG/GME-Workshop (Herausgeber: M. Glesner) 82–91.

[2] ACSL (1999), http://www.acslsim.com.

[3] Adamski, D., Schuster, C. and Hiller, M. (1997), Fahrdynamiksimulation mit FASIM_C++ als Beispiel für die Modellierung mechatronischer Systeme, *Mechatronik im Maschinen- und Fahrzeugbau, VDI Berichte* Nr. 1315, 117–41.

[4] Agsteiner, K., Monjau, D. and Schulze, S. (1996), Objektorientierte Spezifikation digitaler Systeme, 2. GI/ITG/GME-Workshop Hardwarebeschreibungssprachen und Modellierungsparadigmen (Herausgebar: M. Glesner) 128–37.

[5] Ait-Yahia, A., Zerhouni, N., Elmoudni, A. and Ferney, M. (1995), On the modelling and simulation of variable speed continuous Petri nets by design/CPN, *Proc. EUROSIM '95*, 1181–86.

[6] Ajluni, C. (1996), Microsensors Move into Biomedical Applications, *Electronic Design* **44**, 75–84.

[7] Almsick, W. van, Drabe, T., Daehn, W. and Müller-Schloer, C. (1998), A central control engine for an open and hybrid simulation environment, *Proc. IEEE Int. Workshop on Distributed Interactive Simulation and real Time Applications*, 15–22.

[8] Aluru, N. and White, J. (1997), Algorithms for coupled domain mems simulation, *IEEE/ACM Design Automation Conference*, 42.4.

[9] Anantharaman, M. and Hiller, M. (1993), Dynamic analysis of complex multibody systems using methods for differential-algebraic equations, *Advanced Multibody System Dynamics*, (Herausgeber: W. Schiehlen) 173–94.

[10] Anantharaman, M. and Hiller, M. (1996), Integrierte Entwicklungsumgebung zum Entwurf mechatronischer Systeme, *Mess- und Automatisierungstechnik*, VDI-Berichte **1282**, 317–26.

[11] Ansel, Y., Romanowicz, B., Renaud, P. and Schröpfer, G. (1998), Global model generation for a capacitive silicon accelerometer by finite-element analysis, *Sensors and Actuators* **A67**, 153–8.

Mechatronic Systems   Georg Pelz
© 2003 John Wiley & Sons, Ltd   ISBN: 0-470-84979-7

[12]   Antao, B. A. A. (1996), AHD languages — a must for time-critical designs, *IEEE Circuits and Devices Magazine*, **7/96**, 12–17.

[13]   Antón, J., Sandmaier, H., Steger, U., Samitier, J. and Domínguez, C. (1990), Development and simulation of a silicon-based pressure sensor with digital output, *Micromechanics Europe*, no number.

[14]   Ashar K. G. (1997), *Magnetic Disk Drive Technology*, IEEE Press.

[15]   Ashenden, P. J. (1995), *The Designer's Guide to VHDL*, Morgan Kaufmann Publishers.

[16]   Bakalar, K., Christen, E. and Vachoux, A. (1997), Analog and mixed-signal extensions to VHDL through examples, VHDL-AMS Tutorial, `http://www.eda.org/analog`.

[17]   Baltes, H. (1990), Design and simulation of sensors, International Conference on Micro System Technologies **1**, 117–24.

[18]   Barthel, T., Müller, D. and Pauliuk, J. (1999), Spezifikationserfassung für heterogene Systeme, *Tagung Multi-Nature-Systems '99*, Jena, 73–82.

[19]   Bathe, K.-J. (1990), *Finite-Elemente-Methoden*, Springer-Verlag, Berlin, Heidelberg.

[20]   Bechtold, M., Leyendecker, Th., Niemeyer, M., Oczko, A. and Oczko, C. (1990), Das Simulatorkoppelprojekt, *Informatik-Fachberichte* 1990 /255, 244–62.

[21]   Becker, D., Singh, R. K. and Tell, S. G. (1992), An engineering environment for hardware/software co-simulation, *ACM/IEEE Design Automation Conference*, 129–34.

[22]   Bedrosian, G. (1993), A new method for coupling finite element field solutions with external circuits and kinematics, *IEEE Transactions on Magnetics* **29** (1993) /2, 1664–68.

[23]   Beeby, S. P. and Tudor, M. J. (1995), Modelling and optimization of micromachined silicon resonators, *Journal on Micromechanics and Microengineering*, Vol. **5**, 103–5.

[24]   Belli, F. and Dreyer, J. (1994), Systems modelling and simulation by means of predicate/transition nets and logic programming, *Proc. 7th Int. Conf. on Industrial and Engineering Applications of Artificial and Expert Systems*, 465–74.

[25]   Bender, K. and Kaiser, O. (1995), Softwaretest durch Emulation intelligenter Maschinen, 2. Magdeburger Maschinenbau-Tage, Tagungsband II, 1–14.

[26]   Berg, E. C., Lo, N. R., Simon, J. N., Lee, H. J. and Pister, K. S. (1996), Synthesis and simulation for MEMS design, ACM/Sigda Physical Design Workshop, 67–70.

[27]   Berger, M., Jores, P., Fischer-Binder, J.-O. und Staa, P. van (1995), Wo steht Analog-CAE heute? Eine Bestandsaufnahme, GME-Fachbericht 15, *Mikroelektronik*, VDE-Verlag, 259–64.

[28] Bielefeld, J. (1997), Simulation analoger elektromechanischer Mikrosysteme und des Verhaltens von elektrothermischen Bauelementen unter Verwendung eines automatisch generierten vereinheitlichten Modells, Dissertation an der Gerhard-Mercator-Universität-Gesamthochschule-Duisburg.

[29] Bielefeld, J., Gehner, A. and Kunze, D. (1994), Simulation elektromechanischer Verkopplungen am beispiel von mikromechanischen Spiegelelementen, CAD-FEM Users Meeting.

[30] Bielefeld, J., Pelz, G. and Zimmer, G. (1995), Analog hardware description languages for modeling and simulation of microsystems and mechatronics, *Conference on Mechatronics and Robotics 3*, 85–92.

[31] Bielefeld, J., Pelz, G. and Zimmer, G. (1995), Comparison of electrical device representations of physical different equations, *International Conference on Simulation and Design of Microsystems and Microstructures (MicroSIM)* **1**, 267–71.

[32] Bielefeld, J., Pelz, G. and Zimmer, G. (1996), Theoretical foundations of the model transformation approach, *Micro System Technologies*, 133–8.

[33] Bielefeld, J., Pelz, G. and Zimmer, G. (1997), AHDL-model of a 2D mechanical finite-element usable for micro-electro-mechanical systems, *IEEE/VIUF Workshop on Behavioral Modeling and Simulation (BMAS)*, Washington D.C., 177–82.

[34] Bielefeld, J., Pelz, G. and Zimmer, G. (1997), Electrical network formulations of mechanical finite-element models, *2nd International Conference on Simulation and Design of Microsystems and Microstructures, Lausanne*, 239–41.

[35] Bielefeld, J., Pelz, G., Abel, H.-B. and Zimmer, G. (1994), An SOI MOSFET model for circuit simulators considering nonlinar dynamic self-heating, *Proc. IEEE Int. SOI Conference*, 9–10.

[36] Bielefeld, J., Pelz, G., Abel, H.-B. and Zimmer, G. (1995), Dynamic spice-simulation of the electrothermal behaviour of SOI MOSFET's, *IEEE Transactions on Electron Devices*, Vol. **42**, No. 11, Nov. 1995, 1968–74.

[37] Bitzer, R., Deuble, P., Hellmann, M., Schwarz, F. and Wieja, T. (1994), Dezentral Regeln — mit nur einem Chip, *Elektronik* **43**, 70–6.

[38] Bortolazzi, J. and Müller-Glaser, K. D. (1991), Framework integration of an environment for microsystem design, *International Conference on Micro System Technologies*, 153–60.

[39] Bosch (1991), Kraftfahrtechnisches Taschenbuch, 22. Auflage, VDI-Verlag.

[40] Bosch (1992), Autoelektrik und Autoelektronik am Ottomotor, VDI-Verlag.

[41] Bota, S. A., Montané, E., Carmona, M., Marco, S. and Samitier, J. (1997), Parameter extraction scheme for silicon pressure sensors in standard CMOS technology, *International Conference on Simulation and Design of Microsystems and Microstructures (MicroSIM)* **2**, 219–27.

[42] Both, A. W., Biermann, B., Lerch, R., Manoli, Y. and Sievert K. (1994), Hardware-software-codesign of application specific microcontrolers with the ASM environment, *Proc. EuroDAC*, 72–6.

[43] Bradley, E. and Stolle, R. (1996), Automatic construction of accurate models of physical systems", *Annals of Mathematics and Artificial Intelligence* **17**, 1–28.

[44] Bradley, R., Padfield, G. D., Murray-Smith, D. J. and Thomson, D. G. (1990), Validation of helicopter mathematical models, *Transactions Institute of Measurement and Control*, **12**, 186–96.

[45] Brenan, K. E., Campbell, S. L. and Petzold, L. R. (1989), *Numerical Solution of Initial-Value Problems in Differential-Algebraic Equations*, Elsevier Science Publishing.

[46] Brielmann, M. and Kleinjohann, B. (1993), *A Formal Model for Coupling Computer Based Systems and Physical Systems*, EURO-DAC, 158–63.

[47] Brielmann, M. (1995), Modelling differential equations by basic information technology means, *Proc. EUROCAST '95*, 163–74.

[48] Brielmann, M., Stroop, J., Honekamp, U. and Wältermann, P. (1997), Simulation of hybrid mechatronic systems: a case study, *Proc. Int. Conference on Engineering of Computer-Based Systems*, 256–62.

[49] Brockherde, W., Hammerschmidt, D. and Hosticka, B. J. (1995), Silicon microsystems for mechatronic applications, *Conference on Mechatronics and Robotics 3*, 446–55.

[50] Broenink, J. F. (1999), Object-oriented modeling with bond graphs and Modelica, *Proc. International Conference on Bond Graph Modeling and Simulation (ICBGM '99)*, as part of the Western MultiConference.

[51] Buchenrieder, K. and Rozenblit, J. W. (1994), Codesign: an overview in: *Codesign Computer-Aided Software/Hardware Engineering*, IEEE Press 1994, 1–15 (alternative: Proc. 1st Int. Workshop on Software/Hardware Codesign, 1992).

[52] Buck, J., Ha, S., Lee, E. A. and Messerschmitt, D. G. (1994), Ptolemy: A framework for simulating and prototyping heterogeneous systems, *International Journal in Computer Simulation* **4**, 155–82.

[53] Burnett, D. S. (1987), *Finite Element Analysis*, Addison-Wesley Publishing Company.

[54] Butterfield, M. H. (1990), A method of quantitative validation based on model distortion, *Transactions Institute of Measurement and Control*, **12**, 167–73.

[55] Butterfield, M. H. and Thomas, P. J. (1986), Methods of quatitative validation for dynamic simulation models, Part 1: theory, *Transactions Institute of Measurement and Control*, 182–200.

[56] Butterfield, M. H. and Thomas, P. J. (1986), Methods of quatitative validation for dynamic simulation models, Part 2: examples, *Transactions Institute of Measurement and Control*, 201–19.

[57] Cai, X., Osterberg, P., Yie, H., Gilbert, J., Senturia, S. and White, J. (1994), Self-consistent electromechanical analysis of complex 3-D microelectromechanical structures using relaxation/multipole-accelerated method, *Sensors and Materials* **6**, 85–99.

[58] Cameron, R. G. (1992), Model validation by the distortion method: linear state space systems, *IEE Proc.-D*, **139**, 296–300.

[59] Camposano, R. (1996), Embedded System Design, in *Design Automation for Embedded Systems*, 1/96, Vol. **1**, No. 1–2, 5–50.

[60] Capparelli, C. and Nguyen, M. (1996), Visualization, validation and verification of computer simulation and design through virtual prototyping, *Simulators International XIII, Simulation Series*, Vol. **28**, No. 2, 139–45.

[61] Carrol, M., (1993), VHDL — panacea or hype, *IEEE Spectrum*, 6/93, 34–7.

[62] Cellier, F. E. (1991), *Continuous System Modeling*, Springer-Verlag.

[63] Cellier, F. E. (1993), Integrated continuous-system modeling and simulations environments, Kapitel 1 in *CAD for Controlsystems*, (Herausgeber: D. A. Linkens), Marcel Dekker Inc., 1–29.

[64] Cendes, Z. and Ashtiani, C. (1990), Computational electromechanics: entering a new age in engineering design, *International Journal of Applied Electromagnetics in Materials 1*, **2–4**, 281–91.

[65] Chandy, K. M. und Misra, J. (1979), Distributed simulation: a case study in design and verification of distributed programs, *IEEE Transactions on Software Engineering*, Vol. **SE-5**. No. 5, 9/79, 440–52.

[66] Chandy, K. M. and Misra, J. (1981), Asynchronous distributed simulation via a sequence of parallel computations, *Communications of the ACM*, **24** (11), 4/81, 198–206.

[67] Chau, H.-L. and Wise, K. S. (1987), Scaling limits in batch-fabricated silicon pressure sensors, *IEEE Transactions on Electron Devices*, Vol.: **ED-34**, No. 4, April 1987, 850–8.

[68] Choi, B., Lovell, E., Guckel, H., Christenson, T., Skrobis, K. and Kang, J. (1991), Mechanical analysis of pressure transducers with two-sided overload protection, *Journal of Micromechanics and Microengineering*, No. 4, Dec. 1991, 223–229.

[69] Chowanietz, E. G. (1995), Automobile electronics in the 1990s — part 1: powertrain electronics, *Electronics and Communication Engineering Journal*, 23–36.

[70] Chowanietz, E. G. (1995), Automobile electronics in the 1990s — part 2: chassis electronics, *Electronics and Communication Engineering Journal*, 53–8.

[71] Close, C. M. and Frederick, D. K. (1978), *Modeling and Analysis of Dynamic Systems*, Houghton Mifflin Company.

[72] Cobelli, C., Carson, E. R., Finkelstein, L. and Leaning, M. S. (1984), Validation of simple and complex models in physiology and medicine, *American Journal of Physiology* **246**, R259–66.

[73]   Cohen, B. (1995), *VHDL Coding Styles and Methodologies*, Kluwer Academic Publishers.

[74]   Comerford, R. (1994), Mecha... what?, *IEEE Spectrum* **31**, 46–9.

[75]   Crandall, S. H., Karnopp, D. C., Kurtz, E. F. and Pridmore-Brown, D. C. (1968), *Dynamics of Mechanical and Electromechanical Systems*, McGraw-Hill.

[76]   Crary, S. and Kota, S. (1990), Conceptual design of micro-electro-mechanical systems, *International Conference on Micro System Technologies*, 17–22.

[77]   Damm, W., Eckrich, M., Brockmeyer, U., Wittich, G. and Holberg, H. J. (1997), Einsatz formaler Methoden zur Erhöhung der Sicherheit eingebetteter Systeme im Kfz, *Tagung Systemengineering in der KFZ-Entwicklung*, VDI-Bereicht **1374**, 349–66.

[78]   Dammers, D., Binet, P., Pelz, G. and Voßkämper, L. (2001), Motor modeling based on physical effect models, *IEEE International Workshop on Behavioral Modeling and Simulation (BMAS)*.

[79]   Dankert, H. and Dankert, J. (1995), *Technische Mechanik*, B.G.Teubner Verlag.

[80]   Daponte, P., Grimaldi, D. and Savastano, M. (1992), Simulation tools for electrical networks: comparison among existing methodologies, *Proc. EUROSIM '92*, 441–449.

[81]   Deegener, M. and Huss, S. A. (1993), A simulation technique for interacting software — hardware descriptions of automotive systems, *European Simulation Symposium* **5**, 565–70.

[82]   Deiss, H. and Krimmel, H. (1997), Hardware-in-the-Loop Simulation der Spezifikation, Mechatronik im Maschinen- und Fahrzeugbau, VDI-Berichte **1315**, 103–16.

[83]   Donges, E., and Naab, K. (1996), Regelsysteme zur Fahrzeugführung und -stabilisierung in der Automobiltechnik, at *Automatisierungstechnik* **44**, 226–36.

[84]   Donnelly, M. Siegel, C. and Witt, D. (1994), Design analysis of an electronically-controlled hydraulic braking system using the saber simulator, SAE technical paper 940182.

[85]   Dorn, R., Pretschner, A. and Schulze, K.-P. (1996), Ein Beitrag zur Modellierung und Simulation mechatronischer Systeme, *Mess- und Automatisierungstechnik*, VDI-Berichte **1282**, 345–55.

[86]   Dötzel, W. and Billep, D. (1995), Entwurf der Mikromechanischen Komponenten für Sensor- und Aktor-Arrays, me, 24–27.

[87]   Dudaicevs, H., Kandler, M. and Mokwa, W. (1992), Oberflächen-mikromechanik für die Herstellung von Silizium-Drucksensoren, *VDI Berichte — Sensoren: Technologie und Anwendung*, 161–6.

[88]   Dudaicevs, H., Manoli, Y., Mokwa, W., Schmidt, M. and Spiegel, E. (1995), A fully integrated surface micromachined pressure sensor with low temperature dependence, *Proc. Transducers*, 616–9.

[89]   Duttlinger, G. and Filsinger, R. (1990), Adaptives Dämpfungssystem, *Elektronik im Kraftfahrzeug*, VDI-Berichte Nr. 819, no numbers.

[90]   Dymola (1999), `http://www.dynasim.se`.

[91]   Ebert, C. (1998), Experiences with colored predicate-transition nets for specifying and prototyping embedded systems, *IEEE Transactions on Systems, Man and Cybernetics — Part B: Cybernetics*, Vol. **28**, No. 5, 10/98, 641–52.

[92]   Ecker, W. (1993), Using VHDL for HW/SW co-specification, *Proc. Euro-DAC* 1993, 500–5.

[93]   Ecker, W. and Mrva, M. (1996), Objektorientierung: Modellierungs-und Entwurfsparadigma des Jahres 2000, 2. GI/ITG/GME-Workshop Hardwarebeschrei bungssprachen und Modellierungsparadigmen, (Herausgeber: M. Glesner) 118–127.

[94]   Elmqvist, H. and Mattson, S. E. (1997), Modelica — the next generation modeling language, *Proc. World Congress on System Simulation* (WCSS '97).

[95]   Enderlein, V., Härtel, T., Keil, A. and Obermüller, J. (1997), Modellierung und Simulation mechanischer Komponenten von mechatronischen Systemen, *System-Engineering in der KFZ-Entwicklung*, VDI-Berichte **1374**, 479–97.

[96]   Enge, O., Freudenberg, H., Kielau, G. and Maißer, P. (1997), Modellierung und Simulation eines integrierten elektromechanischen Mehrkoordinatenantriebs, *Mechatronik im Maschinen- und Fahrzeugbau*, VDI-Berichte **1315**, 15–25.

[97]   Ernst, R. and Henkel, J. (1992), Hardware-software codesign of embedded controllers based on hardware extraction, *Proc. of the International Workshop on Hardware/Software Codesign*, 1–14.

[98]   Ernst, R. (1999), Automatisierter Entwurf eingebetteter Systeme, at *Automatisierungstechnik*, **7/99**, 285–94.

[99]   Fabula, T. (1992), Finite-Elemente - Modellierung in der Mikromechanik, me 6, II-III.

[100]   Fardoun, A. A., Fuchs, E. F. and Huang, H. (1992), Modelling and simulation of an electronically commutated permanent-magnet machine drive system using SPICE, *IEEE Industry Application Society Conference* (1992), 439–47.

[101]   Fasol, K. H. (1990), Aspekte des Rechnereinsatzes in der Regelungstechnik — eine Bestandsaufnahme, *6. Symposium Simulationstechnik* (ASIM), 13–26.

[102]   Fedder, G. K. and Clark, K. H. (1995), Modelling and simulation of microresonators with meander suspensions, *International Conference on Simulation and Design of Microsystems and Microstructures (MicroSIM)* **1**, 175–83.

[103]   Fedder, G. K. and Jing, Q. (1998), NODAS 1.3 — nodal design of actuactors and sensors, IEEE/VIUF Workshop on Behavioral Modeling and Simulation (BMAS), Orlando.

[104] Folkmer, B., Offereins, H. L., Sandmaier, H., Lang, W., Seidl, A., Groth, P. and Pressmar, R. (1992), A simulation tool for mechanical sensor design, *Sensors and Actuators* **A32**, 521–4.

[105] Frese, L., Neubauer, M. and Ohsendoth-Haase, Ch. (1990), Simulation hybrider Systeme durch gekoppelte Simulatoren, *6. Symposium Simulationstechnik (ASIM)*, 223–7.

[106] Fuchs, M. and Nazareth, D. (1997), Prozessinnovation Steuergeräteentwicklung — ein BMW Experiment basierend auf ASCET-SD, *Systemengineering in der KFZ-Entwicklung*, VDI-Berichte **1374**, 239–50.

[107] Fujimoto, R. M. (1990), Parallel discrete event simulation, *Communications of the ACM* **30** (10), 30–53.

[108] Funk, J. M., Korvink, J. G., Bühler, M., Bächtold, M. and Baltes, H. (1997), SOLIDIS: a tool for microactuator simulation in 3-D, *Journal of Microelectromechanical Systems*, Vol. **6**, No. 1, März '97, 70–82.

[109] Fuss, H. (1990), Wie lang ist der Meter-Stab? — Reflexionen über Daten in Simulations-Modellen —, *6. Symposium Simulationstechnik (ASIM)*, 102–06.

[110] Garthe, D. and Süße, H. (1996), Synthetisierbare VHDL-Modelle zur Verarbeitung ansynchroner Ereignisse, *Hardwarebeschreibungssprachen und Modellierungsparadigmen: 2. GI/ITG/GME-Workshop*, (Herausgeber: M. Glesner) 50–59.

[111] Garverick, S. L. and Mehregany, M. (1996), Methodology for integrated MEMS designs, IEEE Symposium on Circuits and Systems, *IEEE Proc. ISCAS*, Vol. **4**, 1–4.

[112] Gasteier, M. and Glesner, M. (1996), Cosimulation gemischter HW/SW-systeme, Hardwarebeschreibungssprachen und Modellierungsparadigmen, 2. GI/ITG/GME-Workshop, 60–9.

[113] Gasch, R. and Knothe, K. (1989), Strukturdynamik II, Kontinua und ihre Diskretisierung, Springer Verlag.

[114] Gawthrop, P. J. (1993), Symbolic modeling in control, Section 5 in *CAD for Controlsystems* (Herausgeber: D. A. Linkens), Marcel Dekker Inc., 127–46.

[115] Genrich, H. J. and Lautenbach, K. (1981), System modeling with high-level petri nets, *Theoretical Computer Science* **13**, 109–36.

[116] Georgiew, K. (1991), Simulation mit DSPs, *Design und Elektronik*, **24**, 154–5.

[117] Gerlach, G., Schroth, A. and Klein, A. Modellierung nichtelektrischer Komponenten in heterogenen und komplexen Mikrosystemen — Probleme und Lösungswege.

[118] Ghosh, A., Bershteyn, M., Casley, R., Chien, C., Jain, A., Lipsie, M., Tarrodaychik, D. and Yamamoto, O. (1995), A hardware–software co-simulator for embedded system design and debugging, *Proc. of ASP-DAC '95*, 155–64.

[119] Gilbert, J. R., Osterberg, P. M., Harris, R. M., Ouma, D. O., Cai, X., Pfajfer, A., White, J. and Senturia, S. D. (1993), Implementation of a MEMCAD system for electrostatic and mechanical analysis of complex structures

from mask descriptions, *Proc. IEEE Workshop on Micro Electro Mechanical Systems*, 207–12.

[120]   Goel, P. (1992), Rationeller Entwickeln mit HDL — Cadence orientiert sich am Top-Down-Design, *Elektronik* **5/92**, 88–91.

[121]   Götz, A., Krassow, H., Zabala, M. and Cané, C. (1999), CMOS integrated pressure sensor optimization using electrical network simulator-FEM tool coupling, *Journal of Micromechanics and Microengineering*, Vol. **9**, No. 2, 6/99, 109–12.

[122]   Gómez-Cama, J. M., Ruiz, O., Marco, S., López-Villegas, J. M. and Samitier, J. (1997), Simulation of a torsional capacitive accelerometer and interface electronics using an analog hardware description language, *International Conference on Simulation and Design of Microsystems and Microstructures (MicroSIM) 2*, 189–198.

[123]   Gray, G. J. and Murray-Smith, D. J. (1993), The external validation of nonlinear models for helicopter dynamics, *Proc. First Conference of the UK Simulation Society (UKSS '93)*, 143–7.

[124]   Gray, G. J., Li, Y., Murry-Smith, D. J. and Sharman, K. C. (1996), Structural system identification using genetic programming and a block diagram oriented simulation tool, *IEE Electronics Letters*, 7/96, Vol. **32**, No. 15, 1422–24.

[125]   Greenwood, D. T. (1977), *Classical Dynamics*, Prentice Hall.

[126]   Greenwood, D. T. (1988), *Principles of Dynamics, Second Edition*, Prentice Hall.

[127]   Grimm, C. (1996), KIR — ein formales Modell hybrider Systeme, *Hardware-beschrei bungssprachen und Modellierungsparadigmen: 2. GI/ITG/GME-Workshop*, (Herausgeber: M. Glesner) 38–47.

[128]   Grübel, G., Otter, M. and Moormann, D. (1996), Automatisierte Erstellung mechatronischer Simulationsmodelle, *Mess- und Automatisierungstechnik*, VDI-Berichte **1282**, 327–36.

[129]   Günther, M., Denk, G. and Feldmann, U. (1995), Impact of modeling and integration scheme on simulation of MOS-circuits, *Proc. EUROSIM*, 385–90.

[130]   Gupta, R. K., Coelho, C. N. Jr. and De Micheli, G. (1992), Synthesis and simulation of digital systems containing interacting hardware and software components, *IEEE/ACM Design Automation Conference*, 225–30.

[131]   Haase, J., Reitz, S. and Schwarz, P. (1999), Behavioural modeling for heterogeneous systems based on FEM descriptions, IEEE/ACM Workshop on Behavioral Modeling and Simulation (VIUF-BMAS).

[132]   Hale, K. (1994), Automotive database simulation using VHDL, *Proc. Euro-DAC / Euro-VHDL '94*, 592–7.

[133]   Hamam, Y., Rocaries, F. and Carrière, A. (1995), A template for the evaluation of tools for the simulation of continuous system, *Proc. EUROSIM '95*, 165–70.

[134] Heimann, B., Gerth, W. and Popp, K. (1998), *Mechatronik*, Carl Hanser Verlag.

[135] Heinkel, A. (1996), Antriebe perfekt simulieren, *F and M* **104**, 73–5.

[136] Helldörfer, R., Pfeiffer, G., Pressel, J. and Stehr, W. (1997), System-engineering bei der Entwicklung des Telligent-Bremssystems des Actros, *Tagung Systemengineering in der KFZ-Entwicklung*, VDI-Bericht **1374**, 219–36.

[137] Hennecke, D., and Zieglmeier, F. J. (1988), Frequency dependent variable suspension damping — theoretical background and practical success, *Proc. Institution of Mechanical Engineers, Advanced Suspensions*, 101–11.

[138] Hennecke, D., Jordan, B., and Ochner, U. (1987), Elektronische Dämpfer Control — eine vollautomatisch adaptive Dämpfkraftverstellung für den BMW 635 CSi, *ATZ — Automobiltechnische Zeitschrift* **89**, 471–9.

[139] Herbert, D. B. (1992), Simulation and modeling, *IEEE Circuits and Devices* **8**, 11–4.

[140] Herpel, H.-J., Wehn, N. and Glesner, M., Computer-aided prototyping of application-specific embedded controllers in mechatronic systems in: *Codesign Computer-Aided Software/Hardware Engineering, IEEE Press* 1994, 425–42 (alternative: Proc. 1st Int. Workshop on Software/Hardware Codesign 1992).

[141] Herrmann, G. and Müller, D. (1995), Entwurf und Anwendung mikromechanischer Sensor und Aktor-Arrays, me *Mikroelektronik*, 10–12.

[142] Hessel, E. (1995), Model exchange — illusion or future reality?, *Proc. EUROSIM 1995*, 469–74.

[143] Hill, D. D. and Coelho, D. R. (1987), *Multi-level Simulation for VLSI Design*, Kluwer Academic Publishers.

[144] Hiller, M. (1983), *Mechanische Systeme*, Springer Verlag.

[145] Hlupic, V. (1995), A comparison of simulation software packages, *Proc. EUROSIM 1995*, 171–5.

[146] Hockel, K. (1981), Einfluß der zeitlichen Energieumsetzung auf den Wirkungsgrad beim Ottomotor, *Automobil-Industrie*, **3/81**, 315–21.

[147] Hofmann, K. and Glesner, M. (1997), Erstellung und Verifikation von Verhaltensmodellen für Mikrosysteme durch Optimierung, *Mikrosystemtechnik* 1997, 19–32.

[148] Hofmann, K., Glesner, M. Sebe, N., Manolesu, A., Marco, S., Samitier, J., Karam, J.-M. and Courtois, B. (1997), Generation of the HDL-A-model of a micromembrane from its finite-element-description, *European Design and Test Conference*, 108–12.

[149] Hofmann, K., Karam, J.-M., Courtois, B. and Glesner, M. (1997), Generation of HDL-A-Code for nonlinear behavioral models, *IEEE/VIUF Workshop on Behavioral Modeling and Simulation (BMAS)*, 9–16.

[150] Hofmann, K., Karam, J. M., Schulze, M., Theisen, M., Courtois, B. and Glesner, M. (1995), Automatische Übersetzung von FEM-Modellen in eine

analoge Hardwarebeschreibungssprache, *Tagungsband Mikrosystemtechnik und Mikroelektronik*, 86–91.

[151] Hohenberg, G. and Dolt, R. (1993), Ein Konzept zur adaptiven Steuerung/ Regelung von Verbrennungsmotoren unter Verwendung eines Brennver- laufsrechners, *Fachtagunng Integrierte Mechanisch-Elektronische Systeme*, VDI-Fortschritt-Berichte, Reihe 12, Nr. **179**, 58–69.

[152] Honekamp, U. and Kleinschmidt, U. (1997), Der Einsatz von ERCOS in der Softwareentwicklung von Kfz-Steuergeräten, *Systementwicklung in der Kfz-Entwicklung, VDI-Berichte*, Nr. **1374**, 501–12.

[153] Hornbeck, L. J. (1989), Deformable-mirror spatial light modulators, *SPIE Critical Reviews of Optical Science and Technology* **1150**, 86–102.

[154] Horneber, E.-H. (1990), Swith-Level-Timing Simulation digitaler Schal- tungen, *6. Symposium Simulationstechnik (ASIM)*, 556–60.

[155] Huang, N., Cheok, K. C., Horner, T. G. and Settle, T. (1993), Real-time sim- ulation and animation of suspension control system using TI TMS320C30 digital signal processor, *Simulation* **61 : 6**, 405–16.

[156] Hung, E. S., Yang, Y.-J. and Senturia, S. D. (1997), Low-order models for fast dynamical simulation of MEMS microstructures, *Transducers 1997*, 1101–04.

[157] Husinski, I. and Breitenecker, F. (1992), Comparison of simulation soft- ware — the EUROSIM comparisons, *Proc. EUROSIM '92*, 181–86.

[158] IEEE (1988), *IEEE Standard VHDL Language Reference Manual*, IEEE Std 1076-1987, published by IEEE, 345 East 47th Street, New York, NY 10017, USA, (1988).

[159] IEEE (1994), *IEEE Standard VHDL Language Reference Manual*, IEEE Std 1076-1993, published by IEEE, 345 East 47th Street, New York, NY 10017, USA, 1994.

[160] IEEE (1998), *IEEE Standard VHDL Language Reference Manual* (Inte- grated with VHDL-AMS changes), 8/98, http://www.eda.org/analog.

[161] Isermann, R. (1991), Schätzung physikalischer Parameter für dynamische Prozesse 1, at *Automatisierungstechnik 39* **9**, 323–8.

[162] Isermann, R. (1991), Schätzung physikalischer Parameter für dynamische Prozesse 2, at *Automatisierungstechnik 39* **10**, 371–5.

[163] Isermann, R. (1995), Mechatronische Systeme, at *Automatisierungstechnik* **43**, 540–8.

[164] Isermann, R. (1996), On the design and control of mechatronic systems — a survey, *IEEE Transactions on Industrial Electronics*, Vol. **43**, No. 1, 4–15.

[165] Isermann, R. (1996), Mechatronische Systeme — eine Einführung, *GMA- Kongress, Mess- und Automatisierungstechnik*, VDI-Berichte Nr. 1282, 301–5.

[166] Jablonowski, C., Glaser, H., Mäusbacher, B. and Meel, F. van (1997), Entwicklungsablauf einer Fahrdynamikregelung eines AUDI A8 mit Vierradantrieb, *System-Engineering in der KFZ-Entwicklung*, VDI-Berichte **1374**, 115–28.

[167] Jaecklin, V. P., Linder, C., Rooij, N. F. de and Moret, J.-M. (1993), Comb actuators for XY-microstages, *Sensors and Actuators A* **39**, 83–9.

[168] Jefferson, D. R. (1985), Virtual time, *ACM Transactions on Programming Languages and Systems*, Vol. **7**, No. 3, 404–25.

[169] Jefferson, D. R. and Sowizral, H. (1985), Fast concurrent simulation using the time warp mechanism, *Proceedings of the SCS Multiconference on Distributed Simulation, 1985* 63–9.

[170] Jepsen, O. N. (1996), Parametrische Identifikation mit linearen zeitkontinuierlichen Modellen, at *Automatisierungstechnik 44*, **6**, 289–94.

[171] Jorgensen, K. and Odryna, P. (1995), Simulation backplanes allow concurrent use of multiple simulators, *EDN*, **5/95**, 165–72.

[172] Jossen, A. and Späth, V. (1998), Simulation von Batteriesystemen, *Design und Elektronik*, **5/98**, 20–5.

[173] Juang, J.-N. (1994), *Applied System Identification*, Prentice Hall.

[174] Kalavade, A. and Lee, E. A. (1994), Manifestations of heterogenity in hardware/software codesign, *Proc. Design Automation Conference*, 437–38.

[175] Kandler, M., Eichholz, J., Manoli, Y. and Mokwa, W. (1990), CMOS compatible capacitive pressure sensor with read-out electronics, *International Conference on Micro System Technologies*, 574–80.

[176] Kandler, M., Manoli, Y., Mokwa, W., Spiegel, E. and Vogt, H. (1992), A miniature single chip pressure and temperature sensor, MME 1992 — Third European Workshop on Micromachining, Micromechanics and Microsystems, 171–4.

[177] Kane, R. T. and Levinson, D. A. (1980), Formulation of equations of motion for complex spacecraft, *Journal of Guidance and Control*, **3**, 99–112.

[178] Karam, J. M., Courtois, B., Boutamine, H., Drake, P., Poppe, A., Szekely, V., Rencz, M., Hofmann, K. and Glesner, M. (1997), CAD and foundries for microsystems, *IEEE/ACM Design Automation Conference*, 42.2.

[179] Karam, J. M., Courtois, B., Paret, J. M. and Boutamine, H. (1996), Low cost access to MST: manufacturing techniques and related CAD tools, *Micro System Technologies*, 127–32.

[180] Karnopp, D. C. and Rosenberg, R. C., (1975), *System Dynamics: a Unified Approach*, John Wiley & Sons.

[181] Kasper, M., Reill, M. and Weickhmann, M. (1992), Modellierung gekoppelter Systeme, me *Mikroelektronik* 6 (1992)/1, VIII–IX.

[182] Kasper, R., Koch, W., Kayser, A. and Wolf, A. (1997), Integrierte Entwicklungsumgebung mechatronischer KFZ-Komponenten und KFZ-Systeme, *Systemengineering in der KFZ-Entwicklung*, VDI-Berichte **1374**, 451–65.

[183] Kassakian, J. G., Wolf, H.-C., Miller, J. M. and Hurton, C. J. (1996), Automotive electrical systems circa 2005, *IEEE Spectrum*, **8/96**, 22–27.

[184] Kazmierski, T. J., Brown, A. D., Nichols, K. G. and Zwolinski, M. (1991), A general purpose network solving system, *Proc. VLSI '91*, IFIP TC10/WG 10.5, 4a.3.1–4a.3.10.

[185] Kecskeméthy, A. (1993), Objektorientierte Modellierung der Dynamik von Mehrkörpersystemen mit Hilfe von Übertragungselementen, *VDI-Verlag, Reihe 20, Rechnerunterstützte Verfahren, Nr. 88.*

[186] Kecskeméthy, A. (1993), MOBILE — an object-oriented tool-set for the efficient modeling of mechatronic systems, *Second Conference on Mechatronics and Robotics, Moers,* 447–62.

[187] Kemp, T. (1992), A proposal for a modest prototype simulation backplane interface, Revision 1.0, CAD Framework Initiative, Inc.

[188] Keutzer, K. (1994), Hardware-software co-design and ESDA, *Proc. Design Automation Conference,* 435–6.

[189] Kiencke, U., Lacher, F., Schreiber, H. and Ulm, M. (1992), Mikroelektronik im Kraftfahrzeug, neue Anwendungen und Trends — Teil 1, atp *Automatisierungstechnische Praxis* **34**, 231–8.

[190] Kiencke, U., Lacher, F., Schreiber, H. and Ulm, M. (1992), Mikroelektronik im Kraftfahrzeug, neue Anwendungen und Trends — Teil 2, atp *Automatisierungstechnische Praxis* **34**, 314–24.

[191] Kiencke, U., Lacher, F., Schreiber, H. and Ulm, M. (1992), Mikroelektronik im Kraftfahrzeug, neue Anwendungen und Trends — Teil 3, atp *Automatisierungstechnische Praxis* **34**, 375–8.

[192] Kiesewetter, L., Zhang, J.-M., Houdeau, D. and Steckenborn, A. (1992), Determination of Young's moduli of micromechanical thin films using the resonance method, *Sensors and Actuators* **A35**, 153–9.

[193] Kleijnen, J. P. C. (1995), Verification and validation of simulation models, *European Journal of Operational Research* **82**, 145–62.

[194] Klein, A. and Gerlach, G. (1996), System modelling of microsystems containing mechanical bending plates using an advanced network description method, *International Conference on Micro System Technologies,* 299–304.

[195] Klein, A. and Gerlach, G. (1997), Modelling of piezoelectric bimorph structures using an analog hardware description language, *International Conference on Simulation and Design of Microsystems and Microstructures (MicroSIM)* **2**, 229–38.

[196] Klein, A. and Gerlach, G. (1999), Simulation gekoppelter physikalischer Phänomene am Beispiel der Fluid-Struktur-Wechelwirkung einer Mikropumpe, *Multi-Nature-Systems '99,* Jena, 103–12.

[197] Klein, A., Huck, E., Gerlach, G. and Schwarz, P. (1997), Verhaltens- und strukturorientierte Modellierung einer Mikroejektionspumpe, *Mikrosystemtechnik, Chemnitz,* 50–9.

[198] Klein, A., Schroth, A., Blochwitz, T. and Gerlach, G. (1995), Two approaches to coupled simulation of complex microsystems, *Eurosim 1995,* 639–44.

[199] Kleinjohann, B., Tacken, J. and Tahedl, C. (1997), Towards a complete design method for embedded systems using preddicate/transition-nets, *International Conference on Computer Hardware Description Languages and Their Applications,* 4–23.

[200] Klingner, M. and Markert, A. (1993), Logikorientierte Modellierung nicht-linearen Systemverhaltens auf der Grundlage gefärbter Graphen, at *Automatisierungstechnik* **41** (1993) /6 197–204.

[201] Klußmann, J., Krauth, J. and Vöge, M. (1995), Automatic generation of simulation models increases the use of simulation, *Proc. EUROSIM 1995*, 123–8.

[202] Knothe, K. and Wessels, H. (1992), *Finite Elemente*, Springer Verlag.

[203] Kortüm, W. and Troch, I. (1990), Modellreduktion für Simulation und Reglerentwurf, *6. Symposium Simulationstechnik (ASIM)*, 87–91.

[204] Kortüm, W. (1995), Analysis and design of mechatronic vehicles based on MBS codes, *Advances in Automotive Control, IFAC Workshop*, 87–92.

[205] Kortüm, W., Schwartz, W., and Wentscher, H. (1996), Optimierung aktiver Fahrzeugfederungen durch Mechatronik-Simulation, at *Automatisierungstechnik* **44**, 513–21.

[206] Kramer, U. and Neculau, M. (1998), *Simulationstechnik*, Hanser Verlag.

[207] Kreuzer, E. (1979), *Symbolische Berechnung der Bewegungsgleichungen von Mehrkörpersystemen*, Fortschrittberichte VDI, Reihe 11, Nr. 32.

[208] Kreuzer, E. and Schiehlen, W. (1990), NEWEUL — Software for the generation of symbolical equations of motion, (Herausgeber: W. Schiehlen) *Multibody Systems Handbook*, Springer Verlag, 181–202.

[209] Kück, H., Doleschal, W., Gehner, A., Grundke, W., Melcher, R., Paufler, J., Seltmann, R. and Zimmer, G. (1995), Deformable micromirror devices as phase modulating high resolution light valves, *International Conference on Solid-State Sensors and Actuators, and Eurosensors IX* **8**, 301–4.

[210] Kumar, K. B. (1991), Novel techniques to solve sets of coupled differential equations with SPICE, *IEEE Circuits and Devices* **7**, 11–4.

[211] Kunz, W. (1999), private communication.

[212] Kuypers, F. (1993), *Klassische Mechanik*, VCH Verlagsgesellschaft.

[213] Landt, A. (1999), Das Warten hat sich gelohnt — Leica Drive R8, Color Foto 7/99, 22–4.

[214] Lau, K. Y. (1997), MEM's the word for optical beam manipulation, *IEEE Circuits and Devices* **13**, 11–8.

[215] Lautwein, D. (1994), Die Simulation von vollständigen Systemen, *Design und Elektronik* **9**, 59–60.

[216] Lautwein, D. (1995), Mechatronik-Lösung mit Hilfe neuer CAE-Tools, F&M 103/5, Carl Hanser Verlag, 218–20.

[217] Lavagno, L., Chiodo, M., Giusto, P., Jurecska, A., Hsieh, H., Yee, S., Sangiovanni-Vincentelli, A. and Suzuki, K. (1994), A case study in computer-aided codesign of embedded controllers, *International Workshop on Hardware/Software Codesign 3* (1994) 220–4.

[218] Le Marrec, P., Valderrama, C. A., Hessel, F. and Jerraya, A. (1998), Hardware, software and mechanical cosimulation for automotive applications, *Proc. Int. Workshop on Rapid System Prototyping*, 202–6.

[219] Le, T., Renner, F.-M. and Glesner, M. (1997), Hardware in-the-loop simulation — a rapid prototyping approach for designing mechatronics systems, *Proc. 8th IEEE Int. Workshop on Rapid System Prototyping*, 116–21.

[220] Leach, R. (1993), Qualitative modeling in control, in *CAD for Controlsystems*, (Herausgeber: D.A. Linkens), Marcel Dekker Inc., 107–25.

[221] Lee, A. P. and Pisano, A. P. (1993), Repetitive impact testing of micromechanical structures, *Sensors and Actuators* **A39**, 73–82.

[222] Lee, J. L. (1998), *Integriertes Mikrosystem and IC Design, F&M* 106, **11**, 819–22.

[223] Lee, J. S., Yoshimura, S., Yagawa, G. and Shibaike, N. (1995), A CAE system for micromachines: its application to electrostatic micro wobble actuator, *Sensors and Actuators* **A50**, 209–21.

[224] Lee, K.-W. and Wise, K. D. (1982), SENSIM: A simulation program for solid-state pressure sensors, *IEEE Transactions on Electron Devices* ED-29, 34–41.

[225] Lee, S. and Chen, J. (1993), Building models for qualitative prediction of system dynamic behaviour, *International Journal of Pattern Recognition and Artificial Intelligence* **7** (1993) /3, 493–512.

[226] Lefarth, U., Schröer, J. and Siemensmeyer, H. (1990), Simulation als Experiment im Mechatroniklabor, *6. Symposium Simulationstechnik (ASIM)*, 372–6.

[227] Lefarth, U. (1995), SIMEX — an open design environment for modelling and simulating mechatronic systems, *Systems Analysis and Modelling Simulation (SAMS)*, **17**, 27–43.

[228] Leffler, H. (1993), Dynamic stability control DSC — a new BMW control system to improve vehicle stability and handling, *Proceedings of the Institution of Mechanical Engineers*, 83–92.

[229] Lehmann, G. Wunder, B. and Müller-Glaser, K. D. (1996), VYPER! Eine Analyseumgebung zur rechnergestützten Wiederverwendung von VHDL-Entwürfen, *Hardwarebeschreibungssprachen und Modellierungsparadigmen: 2. GI/ITG/GME-Workshop*, (Herausgeber: M. Glesner) 2–14.

[230] Lehmann, G. Wunder, B. and Selz, M. (1994), *Schaltungsdesign mit VHDL*, Franzis Verlag.

[231] Leineweber, M., Pelz, G., Schmidt, M. and Zimmer, G. (2000), New Tactile Sensorchip with Silicone Rubber Cover, *Sensors and Actuators*, A. **84**, 236–245.

[232] Leister, G. and Schiehlen, W. (1990), Modellbildung in der Fahrzeugdynamik, *6. Symposium Simulationstechnik (ASIM)*, 67–71.

[233] Leister, G. and Schiehlen, W. (1990), NEWSIM — ein Simulationswerkzeug für Mehrkörpersysteme, *6. Symposium Simulationstechnik (ASIM)*, 355–9.

[234] Ljung, L. (1993), Identification of linear systems, in *CAD for Controlsystems*, (Herausgeber: D. A. Linkens) Marcel Dekker Inc., 147–65.

[235] Ljung, L., and Glad, T. (1994), *Modeling of Dynamic Systems*, Prentice Hall, Englewood Cliffs.

[236] Lo, N. R., Berg, E. C., Quakkelaar, S. R., Simon, J. N., Tachiki, M., Lee, H.-J. and Pister, K. S. J. (1996), Parametrized layout synthesis, extraction and spice simulation for MEMS, *IEEE Symposium on Circuits and Systems, ISCAS* Vol. **4**, 481–4.

[237] Loftin, R. B. (1995), Training the Hubble space telescope flight team, *IEEE Computer Graphics and Applications*, 9/95, Vol. **15**, No. 5, p. 31–7.

[238] Loftin, R. B. (1995), Virtual environments for aerospace training, *Proc. Northcon 1995*, 31–4.

[239] Long, D. I. and Medhat, S. S. (1992), Simulating mixed-level systems with VHDL, *Proc. EUROSIM '92*, 505–10.

[240] Lorenz, F. (1995), Positioning a standard modeling language, *Proc. EUROSIM 1995*, 487–92.

[241] Lüdecke, A. and Pelz, G., Top-Down Design of a Mechatronic System, *Proceedings: Third Forum on Design Languages (FDL 2000)*, Tübingen, Germany, 151–158.

[242] Lüdecke, A., Trieu, H.-K., Hoffmann, G., Weyand, P. and Pelz, G. (1999), Modeling in hardware description languages for the simulation of coupled fluidic, thermal and electrical effects, *IEEE/ACM Workshop on Behavioral Modeling and Simulation (VIUF-BMAS)*.

[243] Lugner, P. and Bub, W. (1990), Systematik der konzeptionellen Modellbildung, *6. Symposium Simulationstechnik (ASIM)*, 62–6.

[244] Ma, S. (1995), Automatic modeling, simulation and explaining physical systems, *International Conference on Computers and Industrial Engineering 29* (1995) **/1–4**, 187–91.

[245] Magnus, K. and Müller, H. H. (1990), *Grundlagen der technischen Mechanik*, Teubner Verlag.

[246] Maillot, Y. and Wendling, S. (1995), A method for translating automatically statechart models into VHDL code, *Proc. EUROSIM 1995*, 117–22.

[247] Maißer, P. (1988), Analytische Dynamik von Mehrkörpersystemen, *ZAMM Z. angew. Math. Mech.* **68** (1988) 463–81.

[248] Maißer, P. and Steigenberger, J. (1974), Zugang zur Theorie elektromechanischer Systeme mittels klassischer Mechanik — Teil 1: Elektrische Systeme in Ladungsformulierung, *Wissenschaftliche Zeitschrift TH Ilmenau 20*, **6**, 105–23.

[249] Maißer, P. and Steigenberger, J. (1976), Zugang zur Theorie elektromechanischer Systeme mittels klassischer Mechanik — Teil 2: Elektrische Systeme in Flussformulierung, *Wissenschaftliche Zeitschrift TH Ilmenau 22*, **3**, 157–63.

[250] Maißer, P. and Steigenberger, J. (1976), Zugang zur Theorie elektromechanischer Systeme mittels klassischer Mechanik — Teil 3: Elektrische Systeme in gemischter Formulierung, *Wissenschaftliche Zeitschrift TH Ilmenau 22*, **4**, 123–39.

[251] Maißer, P. and Steigenberger, J. (1977), Zugang zur Theorie elektromechanischer Systeme mittels klassischer Mechanik — Teil 4: Elektromechanische Systeme, *Wissenschaftliche Zeitschrift TH Ilmenau* 23, **6**, 151–72.

[252] Maißer, P. and Steigenberger, J. (1979), Lagrange-Formalismus für diskrete elektromechanische Systeme, *ZAMM — Z. angew. Math. Mech.* **59**, 717–30.

[253] Maißer, P. (1997), Alaska — a powerful softwaretool for the investigation of the dynamical behaviour of discrete electromechanical systems, *Proc. Control and Configuration Aspects of Mechatronics (CCAM)*, Ilmenau, 33–49.

[254] Makki, A., Dixit, R. and Billings, R. (1995), Top-down design methods using simulation, SAE Paper 950416, SAE Congress, Vehicle Computer Applications: *Vehicle Systems and Driving Simulation*, 117–22.

[255] Maliniak, L. (1994), Backplanes mesh simulators into one environment, *Electronic Design* (1995) 59–69.

[256] Mallog, J. and Klüting, M. (1989), Einsatz moderner Messverfahren zur Analyse und Optimierung der Ottomotorischen Verbrennung, *MTZ — Motortechnische Zeitschrift* **50**, 275–9.

[257] de Man, H. (1990), Microsystems: a challenge for CAD development, *International Conference on Micro System Technologies*, 3–8.

[258] Mann, H., van Brussel, H. and Yli-Pietilä, T. (1993), Physical level modeling and simulation of multidisciplinary systems, *Proc. 35th SIMS Simulation Conference, Scandinavian Simulation Society*, 51–60.

[259] Mantooth, H. A. and Fiegenbaum, M. (1995), *Modeling with an Analog Hardware Description Language*, Kluwer Academic Publishers.

[260] Marsal, D. (1989), *Finite Differenzen und Elemente*, Springer Verlag.

[261] Mattsson, S. E., Andersson, M. and Åström, K. J. (1993), Object-oriented modeling and simulation, in *CAD for Controlsystems*, D.A. Linkens, Marcel Dekker Inc., 31–69.

[262] Mayeda, W. (1972), *Graph Theory*, John Wiley & Sons.

[263] Mayr, M. and Thalhofer, U. (1993), *Numerische Lösungsverfahren in der Praxis, FEM — BEM — FDM*, Carl Hanser Verlag.

[264] Mazor, M. and Langstraat, P. (1993), *A Guide to VHDL*, Kluwer Academic Publishers.

[265] McCalla, W. (1988), *Fundamentals of Computer-Aided Circuit Simulation*, Kluwer Academic Publishers.

[266] Mee C. D. and Daniel E. D. (1996), *Magnetic Storage Handbook*, 2nd edition, McGraw-Hill.

[267] Mehner, J., Kurth, S., Billep, D., Kaufmann, C., Kehr, K. and Dötzel, W. (1998), Simulation of gas damping in microstructures with nontrivial geometries, *IEEE Int. Workshop on MEMS*, 172–7.

[268] Meisel, J. (1966), *Principles of Electromechanical Energy Conversion*, McGraw-Hill Book Company.

[269] Mikkola, M. (1998), Simulation of diesel electric propulsion systems, *Proc. Diesel Electric Propulsion Conference, Helsinki* (ohne Zählung).

[270] Miu, D. K. (1993), *Mechatronics*, Springer-Verlag.

[271] Mochel, T., Oberweis, A. and Stucky, W. (1992), An open simulation environment for the validation of embedded system designs, *Proc. EUROSIM 1992*, 143–8.

[272] Modelica (1999), Modelica™ — A unified Object-Oriented Language for Physical Systems Modeling, Language Specification,
http://www.modelica.org.

[273] Modelica (1999), Modelica™ — A unified Object-Oriented Language for Physical Systems Modeling, Tutorial and Rationale,
http://www.modelica.org.

[274] Mokwa, W. (1992), Silicon technologies for sensor fabrication, *Chemical Sensor Technology* **4**, 43–62.

[275] Moser, E. (1995), Is VHDL-A suitable as unified modelling language?, *Proc. EUROSIM 1995*, 481–6.

[276] Moser, E. (1996), Automotive model exchange using VHDL-AMS — a benchmark, *Proc. IEEE Int. Symposium on Computer-Aided Control System Design*, 276–81.

[277] Moser, E. and Mittwollen, N. (1998), VHDL-AMS: The Missing Link in System Design — Experiments with Unified Modelling in Automotive Engineering.

[278] Mrčarica, Ž., Glozić, D., Litovski, V. and Detter, H. (1995), Simulation of microsystems using a behavioural hybrid simulator ALECSIS, *International Conference on Simulation and Design of Microsystems and Microstructures (MicroSIM)* 1 (1995), 129–36.

[279] Mrčarica, Ž., Ilić, T., Glozić, D., Litovski, V. and Detter, H. (1995), Mechatronic simulation using Alecsis. Anatomy of the simulator, *Proc. EUROSIM 1995*, 651–6.

[280] Mrčarica, Ž., Litovski, V. B., Delic, N. and Detter, H. (1996), Modelling of micromechanical devices using hardware description language, *International Conference on Micro System Technologies*, 293–8.

[281] Mrčarica, Ž., Randjelovic, Z., Jakovljevic, M., Litovski, V. B. and Detter, H. (1997), Methods for description of microelectro-mechanical device models for system-level simulation, *International Conference on Simulation and Design of Microsystems and Microstructures (MicroSIM)* **2**, 271–80.

[282] Mukherjee, T. and Fedder, G. K. (1997), Structured design of microelectromechanical systems, *IEEE/ACM Design Automation Conference*, 42.3.

[283] Mullen, R. L., Mehregany, M., Omar, M. P. and Ko, W. H. (1991), *IEEE Int. Workshop on MEMS*, 154–9.

[284] Müller, C., Wintz, S. and Rake, H. (1996), Hierarchische Analyse technischer Systeme mit Petri-Netzen, *GMA-Kongress, Mess- und Automatisierungstechnik*, VDI-Berichte Nr. 1282, 47–56.

[285] Müller-Glaser, K. D. (1994), Die goldene Mitte, *Elektronik* **21**, 84–96.

[286] Müller-Glaser, K. D., Rauch, H., Wolz, W., Bortolazzi, J., Kuntzsch, C. and Zippelius, R. (1990), A specification driven environment for microsystem design, *International Conference on Micro System Technologies*, 9–16.

[287] Murray-Smith, D. J. (1995), *Continuous System Simulation*, Chapman & Hall.

[288] Murray-Smith, D. J. (1990), A review of methods for the validation of continuous system simulation models, *Proc. UKSC Conf. on Computer Simulation*, 108–11.

[289] Murray-Smith, D. J. (1995), Advances in simulation model validation: theory, software and applications, *EUROSIM 1995*, 75–84.

[290] Murray-Smith, D. J., Bradley, R. and Leith, D. (1992), Identifiability analysis and experimental design for simulation model validation, *EUROSIM 1992*, 15–20.

[291] Mutz, M. (1996), Korrektheit von high-level Syntheseergebnissen im VHDL-basierten Schaltungsentwurf, *Hardwarebeschreibungssprachen und Modellierungsparadigmen: 2. GI/ITG/GME-Workshop*, (Herausgeber: M. Glesner) 28–37.

[292] Nagel, P., Scharf, R., Wolz, W. and Müller-Glaser, K. D. (1991), A generation environment for simulation models for micro system components, *International Conference on Micro Electro, Opto, Mechanic Systems and Components* **2** (1991) 233–40.

[293] Nair, S. S. (1992), Modeling and simulation of a six-legged walking robot power system, *Simulation* (1992) / March, 185–95.

[294] Nebel, W. and Schumacher, G. (1996), Konzepte objektorientierter Hardwaremodellierung, *2. GI/ITG/GME-Workshop Hardwarebeschreibungssprachen und Modellierungsparadigmen*, (Herausgeber: M. Glesner) 104–5.

[295] Negretto, U. (1991), Predicate/transition nets in FMS control, *Proc. Int. Conference on Industrial Electronics, Control and Instrumentation (IECON)*, 866–71.

[296] Niederhagen, R. (1996), Logiksimulation heute — Werkzeuge zur Verifikation und Validierung, *Elektronik* **9**, 64–70.

[297] Nielan, P. E. and Kane, T. R. (1985), Symbolic generation of efficient simulation / control routines for multibody simulation, *Dynamics of Multibody Systems, IUTAM/IFToMM Symposium*, 153–64.

[298] Niemeyer, M. (1991), Das Simulatorkopplungs-System SiCS, *7. Symposium Simulationstechnik (ASIM)*, 62–6.

[299] Nikravesh, P. E. (1988), *Computer-Aided Analysis of Mechanical Systems*, Prentice Hall.

[300] Ogata, K. (1978), *System Dynamics*, Prentice-Hall Inc., Englewood Cliffs.

[301] Olcoz, S. and Colom, J. M. (1995), A colored Petri net model of VHDL, *Formal Methods in System Design*, 7, Kluwer Academic Publishers, 101–23.

[302] Olcoz, S., Entrena, L. and Berrojo, L. (1995), A VHDL virtual prototyping technique for mechatronic systems design, *Int. Conference on Recent Advances in Mechatronics*, 761–6.

[303] Olcoz, S., Entrena, L. and Berrojo, L. (1995), VHDL virtual prototyping, *IEEE Int. Workshop on Rapid System Prototyping*, **6/95**, 161–7.

[304] Orlandea, N. V., Chace, M. A. and Calahan, D. A. (1976), A sparsity-oriented approach to the dynamic analysis and design of mechanical systems — parts 1 and 2, *ASME Journal of Engineering for Industry*, 773–84.

[305] Orlandea, N. V. (1986), ADAMS Theory and Application, *Seminar on Advanced Vehicle System Dynamics*, **3**, 121–66.

[306] Osterberg, P. M. and Senturia, S. D. (1995), 'MEMBUILDER': an automated 3D solid model construction program for microelectromechanical structures, *Transducers 1995 / Eurosensors IX*, 21–4.

[307] Ott, D. E. and Wilderotter, T. J. (1995), *A Designer's Guide to VHDL Synthesis*, Kluwer Academic Publishers.

[308] Otter, M. (1999), Objektorientierte Modellierung physikalischer Systeme, Teil 1–7, at *Automatisierungstechnik* **47** 1–5, A1–A28.

[309] Otter, M. and Elmquist, H. (1995), The DSblock model interface for exchanging model components, *Proc. EUROSIM 1995*, Elsevier, 505–10.

[310] Otter, M. and Grübel, G. (1994), Direct physical modeling and automatic code generation for mechatronics simulation, *Second Conference on Mechatronics and Robotics, Moers*, 463–82.

[311] Paap, K.-L. (1992), Modellierung für Multi-Level-Simulation, me 9 (1992) 1 VI–VII.

[312] Paap, K.-L., Dehlwisch, M., Jendges, R. and Klaassen, B. (1993), Modeling mixed system with SPICE3, *IEEE Circuits and Devices Magazine, Simulation and Modeling*, 9/93, Vol. **9**, No. 5, 7–11.

[313] Panreck, K. (1999), Systembeschreibungen zur Modellierung komplexer Systeme, at *Automatisierungstechnik*, **4/99**, 157–64.

[314] Papanuskas, J. (1997), IEEE VHDL 1076.1: Mixed-signal behavioral modeling and verification in view of automotive applications, *VHDL Int. User's Forum*, 252–7.

[315] Paschen, U., Leineweber, M., Amelung, J. and Zimmer, G. (1997), Tactile sensors for heavy load manipulation, *Eurosensors* **11**, 1033–6.

[316] Patterson, A. (1996), Eins plus Eins gibt Eins, Single-Kernel- und Backplane-Architekturen konkurrieren bei der Mischsignal-Simulation, *Elektronik*, **6/96**, 90–4.

[317] Paulini, M., Triftshäuser, G., Willutzki, P. and Langenwalter, J. (1997), Aufbau einer Integrationsplattform für physikalische Simulation, *Tagung Systemengineering in der KFZ-Entwicklung*, VDI-Berichte **1374**, 131–46.

[318] Pedersen, M. Olthuis, W. and Bergveld, P. (1995), On the electromechanical behaviour of thin perforated backplanes in silicon condenser microphones, *Transducers 1995 / Eurosensors IX*, 13–16.

[319] Pellerin, D. and Taylor, D. (1997), *VHDL Made Easy!*, Prentice-Hall.

[320] Pelz, G. (2001), Designing Circuits for Disk Drives, *IEEE Int. Conference on Computer Design (ICCD)*.

[321]   Pelz, G. (2001), The Virtual Disk Drive — Mixed-domain support for disk electronics over the complete life-cycle, *IEEE International Workshop on Behavioral Modeling and Simulation (BMAS)*.

[322]   Pelz, G., Bielefeld, J., Zappe, F. J. and Zimmer, G. (1994), MEXEL: simulation of microsystems in a circuit simulator, *International Conference on Micro Electro, Opto, Mechanical Systems and Components* **4**, 651–7.

[323]   Pelz, G., Bielefeld, J., Zappe, F. J. and Zimmer, G. (1995), Simulating micro-electromechanical systems, *IEEE Circuits and Devices Magazine: Simulation and Modeling*, (Editors: R. Saleh und A. Yang), 10–3.

[324]   Pelz, G., Bielefeld, J. and Zimmer, G. (1995), Model transformation for coupled electro-mechanical simulation in an electronics simulator, *Journal on Microsystem Technologies*, 173–7.

[325]   Pelz, G., Bielefeld, J. and Zimmer, G. (1995), Hardware/Software Kosimulation am Beispiel einer prozessor-geregelten Radaufhängung, Analogy Automotive Seminar.

[326]   Pelz, G., Bielefeld, J., Hess, K. G. and Zimmer, G. (1996), Hardware/software-cosimulation for mechatronic system design, *EURO-DAC (1996)* 246–51.

[327]   Pelz, G., Bielefeld, J., Hess, K. G. and Zimmer, G. (1996), Mechatronic system simulation based on analog and digital hardware description languages, *Proc. Mechatronics* **1996**, 421–6.

[328]   Pelz, G., Bielefeld, J. and Zimmer, G. (1997), Modeling of embedded software for automotive applications, *IEEE/VIUF Workshop on Behavioral Modeling and Simulation*, 123–8.

[329]   Pelz, G. (1997), Modeling and simulation of microsystems, invited talk, Workshop on Productivity in Microsystems Design, European Commission, Brussels, 1997.

[330]   Pelz, G. (1997), *Simulating Microelectromechanical Systems*, invited talk at the Thomas Watson Research Center/IBM, Yorktown Heights, New York.

[331]   Pelz, G., Bielefeld, J. and Zimmer, G. (1998), Virtual prototyping for a camera winder: a case study, IEEE/VIUF International Workshop on Behavioral Modeling and Simulation.

[332]   Pelz, G., Kowalewski, T., Pohlmann, N. and Zimmer, G. (1998), Modeling of a Combustion Engine with Hardware Description Languages, IEEE/VIUF International Workshop on Behavioral Modeling and Simulation.

[333]   Pelz, G., Lüdecke, A., Leineweber, M. and Zimmer, G. (1999), Schaltungssimulation mit FE-Modellen für mikroelektromechanische Systeme, *Informationstechnik und technische Informatik (it+ti), Themenheft Mikrosysteme*, 4/99.

[334]   Perry, D. L. (1994), *VHDL, second edition*, McGraw-Hill Series on Computer Engineering.

[335]   Pestel, E. (1988), *Technische Mechanik*, BI Wissenschaftsverlag.

[336] Petrosjanc, K. O. and Maltcev, P. P. (1994), Mixed electrical-thermal and electrical-mechanical simulation of electromechatronic systems using PSPICE, *EuroDAC/Euro-VHDL*, 110–5.

[337] Pichler, F. (1996), Von der Mikroelektronik zur Mikromechatronik, *Elektronik and Informationstechnik, 113*. Jg., Heft 7/8, 547–52.

[338] Popper, K. R. (1934), *Logik der Forschung*, Mohr-Verlag.

[339] Praehofer, H. (1990), Neue Konzepte für die Simulation von kombiniert diskret-kontinuierlichen Systemen — Ein systemtheoretischer Ansatz, *6. Symposium Simulationstechnik (ASIM)*, 112–6.

[340] Prillwitz, G. (1993), Simulatoren: Vielseitigkeit durch Vielfalt, *Elektronik* **18**, 90–7.

[341] Puers, B., Peeters, E. and Sansen, W. (1989), CAD tools in mechanical sensor design, *Sensors and Actuators* **17**, 423–9.

[342] Rammig, F. J. (1993), Modeling Aspects of System Level Design, *Proc. EURO-DAC*, 534–9.

[343] Rammig, F. J. and Horneber, E. H. (1989), Mehrebenensimulation beim Entwurf von VLSI-Schaltungen, me *Mikroelektronik*, Bd. **3**, Heft 6, 278–83.

[344] Reuther, H. M., Weinmann, M., Fischer, M., von Münch, W. and Aßmus, F. (1996), Modeling electrostatically deflectable microstructures and air damping effects, *Sensors and Materials* **8**, 251–69.

[345] Riedel, A. and Schmidt, A. (1990), Fahrdynamiksimulation mit MESA VERDE, *6. Symposium Simulationstechnik (ASIM)*, 360–4.

[346] Risse, W., Vogel, S., Fink, B., Kecskeméthy, A. and Hiller, M. (1993), Object-oriented simulation of a sensor-guided manipulator, *2. Conference on Mechatronics and Robotics, Moers*, 517–30.

[347] Roberson, R. E. and Schwertassek, R. (1988), *Dynamics of Multibody Systems*, Springer.

[348] Roddeck, W. (1997), *Einführung in die Mechatronik*, B.G. Teubner Verlag.

[349] Roduner, C. and Geering, H. P. (1996), Modellbasierte Mehrgrößen-Regelung eines Ottomotors unter Berücksichtigung der Totzeiten, at *Automatisierungstechnik* **44**, 314–21

[350] Romanowicz, B., Schott, C., Laudon, M., Lerch, P., Renaud, P., Popovic, R. S., Amann, H. P., Boegli, A., Moser, V. and Pellandini, F. (1997), Microsystem modelling using VHDL 1076.1, *International Conference on Simulation and Design of Microsystems and Microstructures (MicroSIM)* **2**, 179–88.

[351] Roppenecker, G. and Wallentowitz, H. (1993), Integration of chassis and traction control systems — what is possible — what makes sense — what is under development, *Vehicle System Dynamics*, **22**, 283–98.

[352] Roppenecker, G. (1994), Fahrzeugdynamik: Grundlagen der Modellierung und Regelung, at *Automatisierungstechnik* **42** (1994), 429–41.

[353] Rosenberg, R. C. and Karnopp, D. C. (1983), *Introduction to Physical System Dynamics*, McGraw-Hill.

[354] Ross, B., Fielden, T., Opsahl, G. and Moriyasu, H. (1993), Digital model extraction from measurements, *Proc. NORTHCON*, 76–80.

[355] Rowson, J. A. (1994), Hardware/software co-simulation, *Proc. Design Automation Conference*, 439–40.

[356] Saleh, R., Jou, S.-J. and Newton, A. R. (1994), *Mixed-Mode Simulation and Analog Multilevel Simulation*, Kluwer Academic Publishers.

[357] Saleh, R. A., Antao, B. A. A. and Singh, J. (1996), Multilevel and mixed-domain simulation of analog circuits and systems, *IEEE Trans. on Computer-Aided Design of Integrated Circuits and Systems*, Vol. **1**, No. 15, Jan. 1996, 68–82.

[358] Sandmaier, H., Offereins, H. L. and Folkmer, B. (1993), CAD tools for micromechanics, *Journal of Micromechanics and Microengineering* **3**, 103–6.

[359] Sax, E. Tanurhan, Y. and Müller-Glaser, K. D. (1995), Integrated design process support with VHDL-A, *Proc. EUROSIM* 1995, 493–8.

[360] Scherber, S. and Müller-Schloer, C. (1999), Entwicklungsumgebung zur Modellierung und Simulation heterogener mechatronischer Systeme, *Proc. Multi Nature Systems*, Jena, 53–62.

[361] Schiehlen W., Editor (1990), *Multibody Systems Handbook*, Springer Verlag.

[362] Schlesinger, S. (as Chairman of the SCS Technical Committee on Model Credibility) (1979), Terminology for model credibility, *Simulation*, **32**, 103–4.

[363] Schmerler, S., Tanurhan, Y. and Müller-Glaser, K. D. (1995), A backplane for mixed-mode cosimulation, *Proc. EUROSIM* 1995, 499–504.

[364] Schmerler, S., Tanurhan, Y. and Müller-Glaser, K. D. (1995), A backplane approach for cosimulation in high-level system specification environments, *Proc. EuroDAC* 1995, 262–7.

[365] Schmidt, A. and Wolz, U. (1987), Nichtlineare räumliche Kinematik von Radaufhängungen, *Automobil-Indistrie* **6/87**, 639–44.

[366] Schmitz, H. and Plöger, M. (1997), Konsequenter Einsatz der Simulation in allen Phasen der Fahrzeugsystementwicklung, *Mechatronik im Maschinen-und Fahrzeugbau*, VDI-Berichte **1315**, 91–102.

[367] Scholliers, J. (1994), Development of an integrated environment for the simulation of multitechnical systems, *Proceedings of the European Simulation Multiconference 1994, ESM'94*, 740–4.

[368] Scholliers, J. and Yli-Pietilä, T. (1995), A SPICE-based library for mechatronic systems, *Proc. IEEE Int. Symposium on Circuits and Systems*, 668–71.

[369] Scholliers, J. and Yli-Pietilä, T. (1995), Simulation of mechatronic systems using analog circuit simulation tools, *Proc. IEEE Int. Conf. on Robotics and Automation*, 2847–52.

[370] Schreckenbach, W., Hey, N., Zielke, D., Friedrich, R., Pritzke, B. and Schulte, S. (1994), Simulation of cross coupled effects in physical sensors,

International Conference on Micro Electro, Opto, Mechanical Systems and Components **4**, 833–42.

[371] Schroth, A., Blochwitz, T. and Gerlach, G. (1995), Simulation of a Complex Sensor System using Coupled Simulation Programs, *Transducers 1995/ Eurosensors IX*, 33–5.

[372] Schumann, A. (1990), Eine Methode der rechnergestützten experimentellen Modellbildung, *6. Symposium Simulationstechnik (ASIM)*, 72–6.

[373] Schumann, A. (1994), Rechnergestützte mathematische Modellbildung mittels Computeralgebra, at *Automatisierungstechnik* **42 1**, 23–33.

[374] Schwarz, P. (1996), Modellierung and Simulation komplexer, heterogener Systeme, *10. Symposium Simulationstechnik (ASIM)*, 211–8.

[375] Schwarzenbach, H. U., Korvink, J. G., Roos, M., Sartoris, G. and Anderheggen, E. (1993), A micro electro mechanical CAD extension for SESES, *Journal of Micromechanics and Microengineering* **3** (1993) 118–22

[376] Schweer, G., Tegel, O., Terlinden, M. and Zimmermann, P. (1997), Digital Mock-Up und Virtual reality — Wege zur innovativen Produktentwicklung bei VOLKSWAGEN, *Tagung Systemengineering in der KFZ-Entwicklung*, VDI-Berichte **1374**, 15–37.

[377] Seidel, R., Barthel, T. Albrecht, J. and Müller, D. (1995), Modellierung komplexer Mikrosysteme mit VHDL-A, me *Mikroelektronik*, Heft 5/95, 14–9.

[378] Senturia, S. D. (1995), CAD for microelectromechanical systems, *Transducers 1995/Eurosensors* **IX**, 5–8.

[379] Senturia, S. D. (1996), The future of microsensor and microactuator design, *Sensors and Actuators* **A56**, 125–7.

[380] Senturia, S. D. (1997), Simulation and Design of Microsystems: A Ten-Year Perspective, *Eurosensors* **XI**, 3–13.

[381] Senturia, S. D., Harris, R. M., Johnson, B. P., Kim, S., Nabors, K., Shulman, M. A. and White, J. K. (1992), A computer-aided design system for microelectromechanical systems (MEMCAD), *Journal of Microelectromechanical Systems* **1**, 3–13.

[382] Sharp, R. S. (1994), The application of multibody computer codes to road vehicle dynamics modeling problems, *Proc. of the Institution of Mechanical Engineers*, Vol. **208**, 55–61.

[383] Sibold, T. (1997), Einsatz von CASE-Tools und Hardware-in-the-Loop-Simulation in der Fahrzeugsystementwicklung, *Systemengineering in der KFZ-Entwicklung*, VDI-Berichte **1374**, 251–6.

[384] Smith, D. S. and Howe, D. (1994), Simulation of the dynamic performance of non-linear electro-mechanical systems, *International Conference on Computation in Electromagnetics* (1994) 24–7.

[385] Smola, M. and Wehn, N. (1996), Einsatz von Hardwarebeschreibungssprachen beim Entwurf komplexer Multimedia-Bausteine, Hardwarebeschreibungssprachen und Modellierungsparadigmen: *2. GI/ITG/ GME-Workshop*, (Herausgeber: M. Glesner) 72–80.

[386] Sniegowski, J. J. (1996), Moving the world with surface micromachining, *Solid State Technology*, 83–90.

[387] Spiegel, E., Kandler, M., Manoli, Y. and Mokwa, W. (1992), A CMOS sensor and signal conversion chip for monitoring arterial blood pressure and temperature, *IEEE Int. Solid-State Circuits Conference*, 126–7.

[388] Standridge, C. R., Laval, D. K. and Reust, J. (1992), Model input management: a case study, *Simulation* 199–208.

[389] Stayner, R. M. (1988), Suspensions for agricultural vehicles, *Proceedings of the Institution of Mechanical Engineers, Advanced Suspensions*, 133–140.

[390] Strauss, W. (1997), Veränderungen des Kunden/Lieferanten-Verhältnisses bei der Elektronikentwicklung, *Systemengineering in der Fahrzeugentwicklung*, VDI Berichte **1374**, 41–50.

[391] Suescun, A., Calleja, J., Imbernón, J. and Tomás Celigüeta (1999), Modeling of Complex Mechanical System in VHDL-AMS, IEEE/ACM Workshop on Behavioral Modeling and Simulation (VIUF-BMAS).

[392] Sung, W. and Ha, S. (1998), Efficient and flexible cosimulation environment for DSP-applications, *IEICE transactions Fundamentals*, Vol. **E81-A**, No. 12, 2605–11.

[393] Swaminathan, H., Spracklen, T. and Mathieu, J.-J. (1996), Aspects of model design and development through simulation software, *Simulators International XIII, Simulation Series*, Vol. **28**, No. 2, 146–51.

[394] Swart, N. R., Bart, S. F., Zaman, M. H., Mariappan, M., Gilbert, J. R. and Murphy, D. (1998), AutoMM: automatic generation of dynamic macromodels for MEMS devices, *IEEE Int. Workshop on MEMS*, 178–83.

[395] Tang, W. C. (1997), Overview of microelectromechanical systems and design processes, *IEEE/ACM Design Automation Conference*, 42.1.

[396] Tavangarian, D. (1990), Modellierung digitaler Schaltkreise für eine analoge Simulation, *6. Symposium Simulationstechnik (ASIM)*, 550–5.

[397] Teegarten, D., Lorenz, G. and Neul, R. (1998), How to model and simulate microgyroscope systems, *IEEE Spectrum*, July, 66–75.

[398] Thoma, J. U. (1990), *Simulation by Bondgraphs*, Springer Verlag.

[399] Thomas, D. E., Adams, J. K. and Schmit H. (1993), A model and methodology for hardware-software codesign, *IEEE Design and Test of Computers*, September, 6–15.

[400] Tiller, M., Newman, C. and Trigui, N. (1999), Behavioral modeling of thermodynamic and other non-electrical systems, *IEEE/ACM Workshop on Behavioral Modelung and Simulation (VIUF-BMAS)*.

[401] Timoshenko, S. P. and Woinowski-Krieger, S. (1959), *Theory of Plates and Shells*, McGraw Hill, 27th printing, (1987).

[402] Todesco, A. R. W. and Meng, T. H. Y. (1996), Symphony: a simulation backplane for parallel mixed-mode co-simulation of VLSI-systems, *Proc. Design Automation Conference*, 149–54.

[403] Trah, H.-P. and Ziegenbein, B. (1996), Drucksensoren in Großserie für Kraftfahrzeuge, *Sensor Report*, 26–28.

[404] Turowski, M., Chen, Z. and Przekwas, A. (1998), Squeeze film behaviors in MEMS for large amplitude motion — 3D simulations and nonlinear circuit/behavioral models, IEEE/VIUF Workshop on Behavioral Modeling and Simulation (BMAS), Orlando.

[405] Unbehauen, H. (1994), *Regelungstechnik I*, Kapitel 9, Vieweg Verlag.

[406] Vachoux, A. and Berge, J.-M. (1995), VHDL-A: analog and mixed-mode extensions to VHDL, *Proc. EUROSIM* 1995, 475–80.

[407] Van Zanten, A. T., Erhardt, R., Bartels, H., Hesselbarth, J., Lutz, A. and Neuwald, W. (1997), Simulation bei der Entwicklung der Bosch-Fahrdynamikregelung, *Mechatronik im Maschinen- und Fahrzeugbau*, VDI Berichte Nr. 1315, 143–66.

[408] Van Zanten, A. T., Erhardt, R., Bartels, H., Hesselbarth, J., Lutz, A. and Neuwald, W. (1997), Simulation bei der Entwicklung der Bosch-Fahrdynamikregelung, *Systemengineering in der Kfz-Entwicklung*, VDI Berichte Nr. 1374, 89–114.

[409] Visweswariah, C., Chadha, R. and Chen, C.-F. (1988), Model development and verification for high level analog blocks, *ACM/IEEE Design Automation Conference*, 376–82.

[410] Vlach, J. and Singhal, K. (1994), *Computer Methods for Circuit Analysis and Design*, 2nd Edition, Chapman & Hall.

[411] Vlach, J., Wojciechowski, J. M. and Opal, A. (1995), Analysis of nonlinear networks with inconsistent initial conditions, *IEEE Transactions. on Circuits and Systems — I: Fundamental Theory and Applications*, Vol. **42**, No. 4, 195–200.

[412] Vogelsong, R. S. (1997), Tradeoffs in analog behavioral model development: managing accuracy and efficiency, *IEEE/VIUF Workshop on Behavioral Modeling and Simulation (BMAS)*, Washington, USA, 33–9.

[413] Vogelsong, R. S. (1998), Analog macro/behavioral modeling techniques, *IEEE/VIUF Workshop on Behavioral Modeling and Simulation (BMAS)*, Orlando, USA.

[414] Voigt, P. and Wachutka, G. (1996), Analysis of micropump operation using HDL-A macromodels, *Proc. 26th European Solid State Device Research Conference (ESSDERC)*, 199–202.

[415] Voigt, P., Schrag, G. and Wachutka, G. (1997), Design and Numerical Analysis of an Electro-Mechanical Microsystem for Material Parameter Extraction, *International Conference on Simulation and Design of Microsystems and Microstructures* (MicroSIM) **2**, 209–18.

[416] Voigt, P., Schrag, G. and Wachutka, G. (1998), Microfluidic system modeling using VHDL-AMS and circuit simulation, *Microelectronics Journal* **29**, 791–7.

[417] Voßkämper, L., Lüdecke, A., Leineweber, M. and Pelz, G. (1999), Electromechanical modeling beyond VHDL-AMS, IEEE/ACM Workshop on Behavioral Modeling and Simulation (VIUF-BMAS).

[418] Voßkämper, L., Schmid, R. and Pelz, G. (2000), Combining Models of Physical Effects for Describing Complex Electromechanical Devices, *IEEE/ACM International Workshop on Behavioral Modeling and Simulation (BMAS)*, Orlando, Florida, USA.

[419] Voßkämper, L., Schmid, R. and Pelz, G., Modeling Micro-Mechanical Structures for System Simulations, *Forum on Design Languages (FDL 2001)*.

[420] Voßkämper, L., Schmid, R. and Pelz, G. (2001), Universelle Modellierung mikromechanischer Strukturen für den Einsatz in Systemsimulation, *ASIM Workshop Modellierung und Simulation technischer Systeme*.

[421] Wallaschek, J. (1995), Modellierung und Simulation als Beitrag zur Verkürzung der Entwicklungszeiten mechatronischer Produkte, *Tagung Simulation in der Praxis*, VDI-Bereich **1215**, 35–50.

[422] Wallentowitz, H. (1989), Geregelte Fahrwerke, *Automobil-Industrie* 6/89, 805–19.

[423] Wallentowitz, H. and Janowitz, T. (1996), Der Stellenwert der Berechnung in Industrie sowie in Lehre und Forschung, *Tagung Berechnung und Simulation im Fahrzeugbau*, VDI-Berichte **1283**, 17–46.

[424] Wang S. X. and Taratorin A. M. (1999), *Magnetic Information Storage Technology*, Academic Press.

[425] Waxman, R. and Saunders, L. (1989), The evaluation of VHDL, *IFIP World Computer Congress*, 11/89, 735–42.

[426] Weisser, M. (1996), Spezifikationen nehmen Gestalt an — ein universelles Rapid-Prototyping-Konzept, *Elektronik* **13**, 82–7.

[427] Weisser, M., Rüger, B. and Geisweid, H.-J. (1995), Richtig in die Gänge kommen — durch Rapid Prototyping schneller mit der Funktionsidee ins Auto, *Elektronik* **24**, 72–80.

[428] Wetsel, G. C. and Strozewski, K. J. (1993), Dynamical model of microscale electromechanical spatial light modulator, *Journal of Appled Physics* **73** (11), June, 7120–4.

[429] Willumeit, H.-P. (1998), *Modelle und Modellierungsverfahren in der Fahrzeugdynamik*, B.G. Teubner Verlag.

[430] Wilton, R. (1997), Developments in and applications of mixed-signal HDL tools, *IEE Colloquium on Mixed-Signal AHDL/VHDL Modeling and Synthesis*, 6/1–6/10.

[431] Yli-Pietilä, T., Huovila, H., Yli-Paunu, P. and Wildenburg, L. (1991), Using analog circuit simulation for rapid prototyping of multitechnical devices, *Proc. Int. AMSE Conference on Modeling and Simulation*, Vol. **2**, 67–77.

[432] Yoshii, Y., Nakajo, A., Abe, H., Ninomiya, K., Miyashita, H., Sakurai, N., Kosuge, M. and Hao, S. (1997), 1 chip integrated software calibrated CMOS pressure sensor with MCU, A/D convertor, D/A convertor, digital communication port, signal conditioning circuit and temperature sensor, *Proc. Transducers* 1485–8.

[433] Younse, J. M. (1993), Mirrors on a chip, *IEEE Spectrum*, 27–31.

[434]   Zeid, A. A. (1990), Simple simulation models for complex systems, *Transactions of the Society of Computer Simulation*, Vol. **6**, No. 4, 241–64.

[435]   Zeigler, B. P. (1976), *Theory of Modeling and Simulation*, John Wiley & Sons.

[436]   Zhang, Y., Crary, S. B. and Wise, K. D. (1990), Pressure sensor design and simulation using the CAEMENS-D module, *Solid-State Sensor and Actuator Workshop*, 32–35.

[437]   Zimmerman, W. R. (1996), Time domain solution to partial differential equations using SPICE, *IEEE Transactions on Education*, Vol. **39**, No. 4, 11/96, 563–73.

[438]   Živojnović, V. and Meyr, H. (1996), Compiled HW/SW co-simulation, *Proc. IEEE/ACM Design Automation Conference*, 690–5.

[439]   Zubert, M., Orlikowski, M., Janicki, M., Wociak, W. and Napierals0ki, A. (1997), Methodology of Analogue Silicon Microsystem Modeling Using Hardware Description Language, *Bulletin of the Polish Academy of Sciences, Technical Sciences*, Vol. **45**, No. 4, 525–37.

[440]   Zwoliński, M., Garagate, C. and Kaźmierski, T. J. (1994), Mixed-mode simulation using the alpha simulation backplane, *Colloquium on Mixed-mode Modeling and Simulation*, 6/1–3.

[441]   Zwoliński, M., Garagate, C., Mrčarica, Ž., Kaźmierski, T. J. and Brown, A. D. (1995), Anatomy of a simulation backplane, *IEE Proceedings-Comput. Digit. Tech.*, Vol. **142**, No. 6, Nov., 377–85.

Appendix

Symbols

Lower case letters

a	Acceleration; parameter to be estimated
$\hat{a}$	Identified parameter
$\bar{a}$	Local acceleration
a_{pe}	Angle of accelerator pedal
a_i	Polynomial coefficient
a_{ig}	Ignition advance angle
b	Damping coefficient, generator voltage constant
b_i	Polynomial coefficients
c_W	Drag coefficient
e_k	Error signal at time k
f	Function
g	Gravity
h	Increment
h_i	Form function i
$i_{ij,L}$, $i_{ij,C}$	Inductive and capacitive component of current between the nodes i, j
k	Spring constant
k_{ij}	Entry i, j of the stiffness matrix
m	Mass
m_{fuel}	Fuel mass
m_{tv}	Air mass via the throttle valve
m_{ij}	Entry i, j of the mass matrix
m_{air}	Air mass
m_{idle}	Air mass via the idle setting
n	Number of measurements
n_k	Interference signal
n_{cs}	Rotation speed of crankshaft
p	Pressure
p_{el}	Electrostatic pressure
p_{i0}, p_{i1}	Load at the nodes 0 and 1

Mechatronic Systems   Georg Pelz
© 2003 John Wiley & Sons, Ltd   ISBN: 0-470-84979-7

p_m	Average pressure in the cylinder
Q	Charge
q_i	Value of a boundary condition; generalised coordinate
r	Rotational degree of freedom; exciting function; radial local variable
s	Natural coordinate
s_i	s-coordinate of a Gauss–Legendre integration point
t	Time; natural coordinate
t_c	Thickness of cavity
t_i	Thickness of the insulator
t_j	t-coordinate of a Gauss–Legendre integration point
t_{cs}	Trigger signal at crankshaft
t_p	Plate thickness
u	Voltage; deflection
u_i	Node displacement in x-direction
u_{ij}	Voltage between the nodes i, j
v	Velocity; deflection
$\bar{v}$	Local velocity
v_i	Node displacement in y-direction
w	Excitation
x	Variable; global coordinate
x_k	Input signal at time k
y	Global coordinate
y_k	Output signal at time k
y_k	Output signal of the identified model at time k
z_k	Measured response signal

Upper case letters

A	Area
B_i	Boundary condition i; flexural strength of element i
C	Capacitance
C_i	Constant i
C_{ij}	Capacitance between the nodes i, j; entry i, j of the capacitance matrix
D	Flexural strength; range of a partial differential equation
E	Modulus of elasticity
F	Force; Property of a circuit
I	Moment of inertia
I, i	Current
J	Moment of inertia
L	Inductance, differential operator
L_{ij}	Inductance between the nodes i, j; entry i, j of the inductance matrix
M	Torque

P	Power; parameter
Q	Quality function
Q_i^I	Generalised inertial force
R	Resistance; residuum; radius
S	Sensitivity
T	Energy
U	Voltage
V_D	Piston-swept volume

Greek letters

α	Ratio at a transmission element; angle of road gradient
α_{ij}	Weighting factor for the Gauss–Legendre integration
$\gamma_{xy}, \gamma_{xz}, \gamma_{yz}$	Shear strains
Δ	Laplace operator
ε_0	Dielectric constant
$\varepsilon_{r,eff}$	Effective, relative dielectric constant
$\varepsilon_{xx}, \varepsilon_{yy}, \varepsilon_{zz}$	Strains
ζ	Variable for the approximation of continuous displacements
η	Efficiency
λ	Fuel mixture ratio
μ_i	Mass occupancy of element i
ν	Poisson's ratio
Π	Elastic potential
ρ	Density
$\tau_{xx}, \tau_{yy}, \tau_{zz},$ $\tau_{xy}, \tau_{xz}, \tau_{yz}$	Stresses
ϕ	Solution of a partial differential equation
ϕ_i	Potential at node i
ψ	Magnetic flux
ω	Angular velocity

Vectors / matrices

a	Acceleration matrix
B	Damping matrix
C	Capacitance matrix; material matrix
F	Force vector
H	Interpolation matrix
I	Inertia tensor
J	Jacobi operator
J̄	Global Jacobi matrix
k	Vector of the generalised, gyroscopic forces
K	Stiffness matrix
L	Inductance matrix

$\mathbf{M}$	Mass matrix; torque vector
$\overline{\mathbf{M}}$	Block diagonal matrix of mass and inertia tensors
$\overline{\mathbf{p}}^e$, $\overline{\mathbf{p}}^r$	Applied force/moment vector, reactive force/moment vector
$\mathbf{p}_i$	Load vector at element I
$\hat{\mathbf{u}}$	Vector of node displacement, element displacement vector
ε	Vector of the element strains
τ	Vector of the element stresses
ω	Rotating velocity vector

Abbreviations

AMS	Analogue mixed-signal
DAE	Differential-algebraic equation system
DSP	Digital signal processor
IEEE	Institute of electrical and electronics engineers
FPGA	Field programmable gate array
HDL	Hardware description language
RAM	Random access memory
SCS	Society of computer simulation
VHDL	VHSIC hardware description language
VHSIC	Very high speed integrated circuit

Registered Trademarks

ADAMS	Mechanical Dynamics, Inc., Ann Arbor, MI, USA
ANSYS	Ansys, Inc., Houston, PA, USA
MAST	Avant! Corporation, Fremont, CA, USA
MATLAB/Simulink	The MathWorks Inc., Natick, MA, USA
$MATRIX_X$	Integrated Systems, Inc., Sunnyvale, CA, USA
Modelica	Modelica Design Group
PSpice	MicroSim Corp., Irvine, CA, USA
Saber	Avant! Corporation, Fremont, CA, USA
Sparc	SUN Microsystems, Palo Alto, CA, USA
Verilog	Cadence Design Systems, Inc., San Jose, CA, USA
VHDL system simulator (VSS)	Synopsys, Inc., Mountain View, CA, USA

Index

Mechatronic Systems Georg Pelz
© 2003 John Wiley & Sons, Ltd ISBN: 0-470-84979-7